A Beginner's Guide to Using Open Access Data

POCKET GUIDES TO
BIOMEDICAL SCIENCES

Series Editor
Lijuan Yuan
Virginia Polytechnic Institute and State University

https://www.crcpress.com/Pocket-Guides-to-Biomedical-Sciences/
bookseries/
CRCPOCGUITOB

The *Pocket Guides to Biomedical Sciences* series is designed to provide a concise, state-of-the-art, and authoritative coverage on topics that are of interest to undergraduate and graduate students of biomedical majors, health professionals with limited time to conduct their own searches, and the general public who are seeking for reliable, trustworthy information in biomedical fields.

A Beginner's Guide to Using Open Access Data

Saif Aldeen Saleh AlRyalat
Shaher Momani

CRC Press
Taylor & Francis Group
Boca Raton London New York

CRC Press is an imprint of the
Taylor & Francis Group, an **informa** business

CRC Press
Taylor & Francis Group
6000 Broken Sound Parkway NW, Suite 300
Boca Raton, FL 33487-2742

International Standard Book Number-13: 978-0-367-07506-4 (Hardback)
978-0-367-07503-3 (Paperback)

Library of Congress Cataloging-in-Publication Data

Names: AlRyalat, Saif Aldeen S., author. | Momani, Shaher M., author.
Title: A beginner's guide to using open access data / Saif Aldeen S. AlRyalat and Shaher M. Momani.
Other titles: Pocket guides to biomedical sciences.
Description: Boca Raton : Taylor & Francis, 2019. | Series: Pocket guides to biomedical sciences | Includes bibliographical references.
Identifiers: LCCN 2018046937| ISBN 9780367075033 (pbk. : alk. paper) | ISBN 9780367075064 (hardback : alk. paper) | ISBN 9780429021060 (general) | ISBN 9780429667671 (pdf) | ISBN 9780429664953 (epub) | ISBN 9780429662232 (mobi/kindle)
Subjects: | MESH: Data Mining | Datasets as Topic | Access to Information | Biomedical Research--methods
Classification: LCC R855.3 | NLM W 26.55.I4 | DDC 610.285--dc23
LC record available at https://lccn.loc.gov/2018046937

Visit the Taylor & Francis Web site at
http://www.taylorandfrancis.com

and the CRC Press Web site at
http://www.crcpress.com

Contents

Series Preface

Dramatic breakthroughs and nonstop discoveries have rendered biomedicine increasingly relevant to everyday life. Keeping pace with all these advances is a daunting task, even for active researchers. There is an obvious demand for succinct reviews and synthetic summaries of biomedical topics for graduate students, undergraduates, faculty, biomedical researchers, medical professionals, science policy makers, and the general public. Recognizing this pressing need, CRC Press established the Pocket Guides to Biomedical Science series, with the main goal being to provide state-of-the-art, authoritative reviews of far ranging subjects using short, readable formats intended for a broad audience. The volumes in this series will address and integrate the principles and concepts of the natural sciences and liberal arts, especially those relating to biomedicine and human well-being. Future volumes will come from biochemistry, bioethics, cell biology, genetics, immunology, microbiology, molecular biology, neuroscience, oncology, parasitology, pathology, and virology, and other related disciplines.

This current volume addresses the use of open-access data as a source for cutting-edge scholarship, especially for biomedical research. Over the past three decades, since the formation of the Internet in 1991, a large number of government agencies, research institutions, and academic institutions across the world have created databases in various fields of science and technology. Today, many of these databases are freely available on the World Wide Web through the individual portals and websites of these organizations. Open-access data, such as clinical trial metadata, genome, proteome, microbiome, and metabolome from many repositories, provide ever-increasing opportunities for using these data for research. This concise volume provides practical guidance for how to use open-access data in research, from generating research ideas to publishing in referred journals. Both young researchers and well-established scholars can use this book to upgrade their skills with respect to data sources, accessing open data, ethical considerations, analyzing data to reach conclusions, writing manuscripts, publishing, and even post-publishing promotion.

The goal of this volume is the same as the goal for the series—to simplify, summarize, and synthesize a complex topic so that readers can reach to the core of the matter without the necessity of carrying out their own time-consuming literature searches. We welcome suggestions and recommendations from readers and members of the biomedical community for future topics in the series, and urge other experts to serve as future volume authors/editors.

Lijuan Yuan
Blacksburg, Virginia, USA

Preface

Medical research has changed the face of healthcare, increased life expectancy by about 20 years, led to the eradication of smallpox, and made cancer a treatable disease. On the other hand, the people who performed such life-changing research had participated in several studies before that. The point is, you need to know the steps of how to do research before asking for funding for your possibly groundbreaking study. But how can you do research without financial backing, or sometimes without proper mentorship, which is the case in most developing countries? This book was written for the purpose of answering this question by guiding early career researchers on how to use what is known as *open access data* in performing research. *Open data* is data that can be utilized freely by others, and in our context, the use is for research purposes.

This book begins with Chapter 1, which provides an overview on open data, their importance for researchers, and some examples of how they were used in research. After that, it provides insight into the open data resources available for researchers and defines two main types: explicitly available open data, which can be downloaded directly by researchers who wish to analyze it, and implicitly available open data, which requires researchers to perform one or more processing steps to get the information. After obtaining the data, you now need to deal with it and generate research ideas. An important step in both generating ideas and performing research is literature review, and this discussion provides tips on the most efficient way of doing this. Now that you have decided that you are interested in a specific data set, it is time to access it, and this chapter explains the entire process of accessing open data. Finally, there are ethical considerations that you need to keep in mind during your research on open data. At the end of this chapter, a tutorial on how to come up with a research project and access the open data that you need to execute it is provided, where you either follow the example provided or generate your own during this book.

Chapter 2 begins by providing an overview of research and the benefits of doing it, in addition to correcting some misconceptions about it. Conducting research properly requires good planning before you can even think about creating a manuscript, so we explain the components of the planning and executing phases during a research study. Knowing the type of study you are doing ensures high-quality output, and this chapter covers the main types of research and an overview for each type. In addition, this discussion provides a prospective overview of the life cycle of a study, so that you will be prepared for the next steps of your research. Finally, research must be done to meet high ethical standards, a topic covered in detail in this chapter. Doing research using open data is slightly different from generating data de novo, so, in this chapter, we stress these differences. At the end of this chapter, there is a tutorial in which we write the protocol for the idea from

Chapter 1. We advise you to create a protocol for the data that you should've accessed in that chapter.

In Chapter 3, you will gain insight into an important aspect of research: statistical analysis. We begin with an overview of the basics of statistical analysis, a task that is essential to researchers, both to interpret articles you read and to analyze your own data. After that, we present a simple diagram to help you analyze your data, which will guide you on choosing the statistical tests that you will need to perform on your hypothesis in the simplest way possible. Although test formulas and complicated statistical analyses are beyond the scope of this book, we present a list of references at the end of the chapter if you would like to learn more on this topic. You will not need a biostatistician to do your first research study, but we recommend that you consult an experienced statistician when you finish your analysis.

Finally, in Chapter 4, we walk you through the steps of writing your manuscript according to the most recent reporting guidelines, where we focus on the points of manuscript writing specific to open data research. As you write your manuscript, we recommend that you go step by step, in the same order presented in this book. After completing your manuscript, you probably will want to publish it. The final section of this chapter provides the information you need to publish in a strong journal that best fits the scope of your study. As professor Momani said, "A great study published in a weak journal will become a weak study, and a weak study published in a strong journal will become a strong study."

Acknowledgments

Dr. Christine Darby thoroughly reviewed the book and provided us with highly valuable comments to improve its contents. Hussein Al-Ryalat designed the figures in this book.

Authors

Dr. Saif Aldeen Saleh AlRyalat is a medical doctor in the Department of Ophthalmology at the University of Jordan in Amman. He is one of the renowned young researchers in Jordan, where he has published in several biomedical and general research disciplines. Dr. AlRyalat was a keynote speaker at several international biomedical conferences. He provided several research courses on biomedical research for medical students, residents, and high-degree specialists, where at the end of the course, students are expected to have published their own studies using open access data. Along with his many publications, he is the first Jordanian patentee on the international Patent Corporation Treaty (PCT). (https://orcid.org/0000-0001-5588-9458)

Professor Shaher Momani is a leading scientific researcher at the University of Jordan in Amman, and has authored or coauthored more than 250 peer-reviewed papers in Institute for Science Information's (ISI) international journals of high quality. Momani was appointed by Thomson Reuters to its prestigious list of Highly Cited Researchers in 2014, 2015, and 2016. He is the only scientist in the Arab world who has been chosen for this prestigious honor for three consecutive years. Momani obtained the highest Hirsch Index (h-index) and number of citations in Jordan, according to the Scopus database and Google Scholar.

In addition to The Order of King Abdullah II Ibn Al Hussein for Excellence of the Second Class, Momani has received many honors and awards, including the Ali Mango Distinguished Researcher Prize in Jordan (2016), the Distinguished Researcher Prize in Jordan (2012), the Distinguished Researcher Prize at the University of Jordan (2012), the Islamic Educational, Scientific, and Cultural Organization Science Prize (ISESCO Science Prize) (2008), the Scopus Prize for Jordan Scientists (2009), the Distinguished Researcher Prize at Mutah University (2009), the TWAS Prizes for Young Scientists in Developing Countries, Third World Academic Sciences (2000), and the Award of Jordan National Commission for Education, Culture, and Science (2008). Momani was nominated for the Nobel Prize in Physics in 2016 by many scholars and institutions throughout the Arab world. (https://orcid.org/0000-0002-6326-8456)

Lna W. Malkawi is a medical doctor in the Department of Radiology and Nuclear Medicine at the University of Jordan in Amman. Dr. Malkawi is a pioneer in the area of ethical research in Jordan, where she provides consultations on biomedical ethics in research and practice. She has several publications in biomedical and open data research.

1

Open Data

Saif Aldeen Saleh AlRyalat

Data overview

In 2013, the world as a whole spent $1.48 trillion on research, with the United States and China alone spending around $1 trillion (ISSC, IDS, and UNESCO, 2016). On the other hand, the world's low- and low-middle-income countries' expenditure on research is exceedingly low, apparently because there are other priorities to spend money on instead (World Bank, 2018). This lack of sufficient research funds resulted in an incorrect belief among developing-world researchers and academics that it is difficult to do research in these regions (Horton, 2000). Moreover, funding bodies usually require researchers requesting monies to have previous experience, and even if researchers are able to get funding, they will struggle to find educated staff to support their work. Thus, a vicious circle begins, of researchers not being able to do studies because they can't get funding, which requires research experience. An example of this situation occurs in Africa, where most universities do not put sufficient emphasis or funding for research, even for academic purposes (Horton, 2000), which results in their staff lacking research experience (Vernon et al., 2018).

Collecting high-quality data for any study is the step that requires the most money and experience, as well as time (Dicks et al., 2014). These data collection fundamentals are not readily available in developing countries, thus hindering research progress in these countries—the aforementioned vicious circle. To reduce the huge costs of research in developing countries, researchers and universities should depend on data collected by other more experienced researchers, who have sufficient funds to support their work. This model of research, based on using others' data to do research, is known as *open-access data research (open data research)*.

Open data research has emerged recently as a new model of research, one that wasn't possible in the past due to lack of resources, systems technology to access data, bioinformatics expertise, and legal infrastructure to facilitate sharing (Bertagnolli et al., 2017). On the other hand, the open data approach is expected to need a good deal of support to reach its potential. To further show the importance of promotion of and support for open data, in 2003, the University of Rochester, one of the most highly ranked universities in New York, launched a digital archive (i.e., a repository) designed to share dissertations, preprints, working papers, photographs, music scores, and any other kind of digital data that the university's investigators could produce.

Six months of research had convinced the university and its investigators that a publicly accessible online archive would be well received. At the time that the repository was launched, the university librarians were worried that the flood of data that would be uploaded might overwhelm the available storage space. Six years later, the $200,000 repository was mostly empty (Nelson, 2009).

To promote open data research, an annual award was created to recognize data sharers whose data have been reused in impactful ways: the Research Symbiont Award (www.researchsymbionts.com). There was also a call for journals to promote open data research by publishing research done using open data, especially those journals already published the primary publication (Byrd, 2017). Moreover, a study that compared citation rates among papers with their data made openly available and those that didn't found that making data openly available increased a paper's citation rates, which further motivates and increases open data research (Piwowar et al., 2007).

Open data may be freely used, reused, and redistributed by anyone (Dietrich et al., 2009). Open data have three main characteristics:

- *Accessible*: The data must be available as a whole, and at no more than a reasonable reproduction cost, preferably by downloading over the Internet. The data must also be available in a convenient and modifiable way, preferably in a machine-learning form (e.g., a Microsoft Excel spreadsheet).
- *Reuse and redistribution*: The data must be provided under terms that permit their reuse and redistribution, including intermixing with other data sets (e.g., combining two data sets on the same topic to create a larger one).
- *Universal participation*: Everyone must be able to use, reuse, and redistribute the data.

The concept of open data is mostly related to the field of economics, where governments and companies can use open data to improve their functionality and products. An example would be manufacturers using data obtained from their products already on the market to improve the development of their next products and to create innovative after-sales offerings (Manyika et al., 2011). The same concept can be used by scientists who use data on healthcare consumers collected by pharmaceutical companies, healthcare institution records, insurance companies, and other local and national authorities to perform research. The following example will further clarify how researchers can benefit from open data.

In 2010, the protocol of a large study called the Systolic Blood Pressure Intervention Trial (SPRINT) was finalized, stating that it would include 9361 patients from 102 centers in the United States, and they would be followed up for eight years (Ambrosius et al., 2014). The total cost of this single study was estimated to be approximately $157 million (NHLBI, n.d.)—more than the available research funds for several countries collectively. The aim of SPRINT was to compare the outcomes of standard versus intensive systolic blood pressure (SBP) lowering. It recruited 4,678 patients to receive intensive SBP lowering (SBP < 120 mm Hg), and another 4,683 patients to receive standard SBP lowering (SBP < 140 mm Hg). At the end of the study, four months later, more than 140 novel articles were written using this single data set by researchers from all over the world, including developing countries, for no additional cost.

The open data model was suggested to allow maximum benefits from research; instead of spending millions of dollars to publish only one study, open data allows new findings and new publications to be derived from that single set of data. Innovations from researchers may be communicated and used in other parts of the world, including developing countries. For this reason, in March 2017, the *New England Journal of Medicine (NEJM)* initiated a challenge for researchers to conduct a novel study from data already collected from the SPRINT study (Burns and Miller, 2017). The winner of the NEJM contest developed a weighted risk–benefit calculator to examine the pros and cons of intensively treating patients with hypertension. The second-place finisher, consisting of medical students, performed an analysis to only part of the population involved in the SPRINT trial and concluded that intensive treatment produced more harm and was more difficult in patients who had chronic kidney disease. The third-place winner developed a risk calculator to help physicians treating patients with hypertension in their decision-making (Burns and Miller, 2017). If the 140+ studies done using SPRINT data had needed to collect fresh data themselves instead, the cost would have been around $140 million, considering that the average cost of obtaining high-quality data on hypertension would be around $1 million (United Kingdrom Research and Innovation Gateway (https://gtr.ukri.org)).

Another example of the open data model was the Framingham Heart Study (Dawber et al., 1951), a cohort study started in 1948 under the direction of the National Heart, Lung, and Blood Institute (NHLBI) with the aim of identifying the common factors or characteristics that contribute to cardiovascular disease. Prior to this study, almost nothing was known about the epidemiology of hypertensive or arteriosclerotic cardiovascular disease. Much of the now-common knowledge concerning heart disease, such as the potential effects of diet, exercise, and common medications such as aspirin on cardiovascular risk, was based on this longitudinal study. The data from this ongoing study are available for future researchers. Now, more than 1,000 medical studies related to the Framingham Heart Study have been published without any data costs to their authors. These studies would have cost millions, and some of them were done by authors from developing countries and early career researchers who would not have obtained sufficient funding to the studies if they had to generate their own data.

These two examples show how using open data is sometimes the perfect way of doing research. The model of using open data to do research is the perfect options if you find yourself in one of three scenarios:

1. You have an idea about what you would like to study, but you do not have the required resources to conduct a study at your institution. In addition, a large open data study has already been done on the same subject in the past. For example, you would like to study Alzheimer's disease but do not have the resources to conduct it; however, you do have access to an open data study about Alzheimer's called the Alzheimer's Disease Neuroimaging Initiative (ADNI).
2. You do not have a specific idea, but you are interested in doing research in a certain specialty in general (e.g., neurology). In this case, you could look for open data resources on neurology and access the one that you think best fits your interests. You would generate your idea from that resource.
3. You would like to train new researchers who had never published articles before. Instead of spending resources on collecting data for teaching purposes, you may simply use open data repositories to train them at minimal cost.

Data available

Generally, there are two main types of open data:

- Explicitly available (e.g., already in the form of downloadable Excel spreadsheets).
- Implicitly available (e.g., needs an extra step to put the data in the form of Excel spreadsheets).

Explicitly available open data

Explicitly available open data can be directly accessed and used, where either you can download it directly in the form of Excel spreadsheets (or any other available software), or you need to gain access to it through the gatekeeper model. This type is unlike implicitly available open data, where you need to do some processing before reaching the analyzable final form (e.g., Excel spreadsheet).

These data are usually provided from countries, institutions, or publically available repositories for individual use. An example of a country-level database is the U.S. Open Data Repository (www.data.gov), where the government provides access to more than 250,000 high-value data sets regarding all government information. At the level of institutions, Yale University launched the YODA Project (http://yoda.yale.edu) to provide data generated from clinical studies for other researchers to use. At the individual level, Figshare (www.figshare.com) allows users to upload files in any format to be previewed in the repository browser.

Although data are widely available, they vary with regard to their quality or the methodology of their collection. You should always check the quality of the data before using them, which you can evaluate using the following tips:

1. *The quality of the repository itself:* For example, data deposited at National Institutes of Health (NIH) repositories already have been checked for quality.
2. *The primary research published using the data:* If the primary research is published in a high-quality journal, its data would be of higher quality compared to research published in a lower-quality journal. (Note: For a guide on checking the quality of a journal, see section 4.2 in Chapter 4.)
3. *The protocol that details how the data was collected:* This is usually deposited with the original data.

As the first step of choosing the data is checking the quality of the repository, we will discuss three high-quality data repositories, which we believe will be your starting point for selecting data to base your research on. First; the U.S.-based NIH is the world's largest single funder of research in the healthcare and biological sciences, with an annual budget of over $30 billion (Li et al., 2017). Almost all data for research projects funded by NIH are high quality and are required to deposit the data in an NIH-supported repository (NIH data-sharing policies). The Trans-NIH BioMedical Informatics Coordinating Committee (BMIC) is responsible for coordinating among NIH centers and providing data from each center on one page.

The second repository is FAIRsharing (www.FAIRsharing.org), which is a searchable, Web-based data portal affiliated with the University of Oxford. Each repository listed on this site is checked to ensure the quality.

The third repository, re3data (https://www.re3data.org), is supported by a nonprofit organization called Science Europe. It ensures that repositories listed within it have persistent identifiers, metadata, data access and use, machine-readability, and long-term preservation. Repositories can be browsed by subject, country, or content type in a user-friendly interface.

In 2014, the Nature Publishing Group started a new journal called *Scientific Data* (www.nature.com/sdata), with the goal of enabling the discoverability, reproducibility, and reuse of valuable data. It contains descriptions of scientifically valuable data sets, including detailed methods and technical analyses attesting to the data quality. However, *Scientific Data* does not host the data sets itself. These data sets must be made available via *recognized repositories*, which are repositories evaluated by a special team to ensure their quality. Table 1.1 contains biological and health sciences data repositories that are recognized by *Scientific Data*.

The median cost of conducting a clinical trial, from protocol approval to final report, is $3.4 million for phase I trials, $8.6 million for phase II trials, and $21.4 million for phase III trials, most of which are funded by either

Table 1.1   Open Data Repositories That Were Quality Checked by _Scientific Data_

Category	Repository	Website
Nucleic acid sequencing	DNA DataBank of Japan (DDBJ)	https://www.ddbj.nig.ac.jp/index-e.html
	European Nucleotide Archive (ENA)	https://www.ebi.ac.uk/ena
	GenBank	https://www.ncbi.nlm.nih.gov/genbank/
	Database of single nucleotide polymorphisms (SNPs)	https://www.ncbi.nlm.nih.gov/snp
	European Variation Archive	https://www.ebi.ac.uk/eva/
	NCBI's database of human genomic structural variation (dbVar)	https://www.ncbi.nlm.nih.gov/dbvar/
	Database of Genomic Variants archive (DGVa)	https://www.ebi.ac.uk/dgva
	EBI Metagenomics	https://www.ebi.ac.uk/metagenomics/
	Trace Archives	https://www.ncbi.nlm.nih.gov/Traces/home/
	Sequence Read Archive (SRA)	https://www.ncbi.nlm.nih.gov/sra
	Assembly	https://www.ncbi.nlm.nih.gov/assembly
Protein sequence and proteomics	UniProtKB	https://www.uniprot.org
	PeptideAtlas	http://www.peptideatlas.org
	PRIDE PRoteomics IDEntifications (PRIDE)	https://www.ebi.ac.uk/pride/archive/
Molecular structure and metabolomics	Protein Circular Dichroism Data Bank	http://pcddb.cryst.bbk.ac.uk/home.php
	Crystallography Open Database	http://www.crystallography.net/cod/
	Biological Magnetic Resonance Data Bank	http://www.bmrb.wisc.edu
	EMDataBank	http://www.emdatabank.org
	Protein Data Bank archive (PDB)	https://www.wwpdb.org
	The SBGrid Data Bank	https://data.sbgrid.org
	Metabolights	https://www.ebi.ac.uk/metabolights/

(Continued)

Table 1.1 (Continued) Open Data Repositories That Were Quality Checked by *Scientific Data*

Category	Repository	Website
Neuroscience	NeuroMorpho.Org	http://neuromorpho.org
	International Neuroimaging Data-Sharing Initiative (INDI)	http://fcon_1000.projects.nitrc.org
	OpenNEURO	https://openneuro.org
	GIN	https://web.gin.g-node.org
Imaging	Coherent X-ray Imaging Data Bank	http://www.cxidb.org
	Image Data Resource	http://idr.openmicroscopy.org/about/
	The Cancer Imaging Archive	http://www.cancerimagingarchive.net
	SICAS Medical Image Repository	http://www.smir.ch
Nonhuman organisms	Eukaryotic Pathogen Database Resources (EuPathDB)	https://eupathdb.org/eupathdb/
	FlyBase	http://flybase.org
	Influenza Research Database	http://www.fludb.org/brc/home.spg?decorator=influenza
	Mouse Genome Informatics (MGI)	http://www.informatics.jax.org
	Rat Genome Database (RGD)	https://rgd.mcw.edu
	VectorBase	https://www.vectorbase.org/index.php
	Xenbase	http://www.xenbase.org/entry/
	Zebrafish Model Organism Database (ZFIN)	http://zfin.org

(*Continued*)

Table 1.1 (*Continued*)   Open Data Repositories That Were Quality Checked by *Scientific Data*

Category	Repository	Website
Health sciences	National Addiction and HIV Data Archive Program (NAHDAP)	http://www.icpsr.umich.edu/icpsrweb/NAHDAP/index.jsp
	National Database for Autism Research (NDAR)	http://ndar.nih.gov
	ClinicalTrials.gov	https://clinicaltrials.gov
	PhysioNet	http://physionet.org
	National Database for Clinical Trials (NDCT) related to mental illness	http://ndct.nimh.nih.gov
	Research Domain Criteria Database (RDoCdb)	http://rdocdb.nimh.nih.gov
	UK Data Service	http://discover.ukdataservice.ac.uk

Note: A more detailed list is available at https://www.nature.com/sdata/policies/repositories#life.

governmental entities or pharmaceutical companies (Martin et al., 2017). These studies provide valuable, high-quality data for subsequent analysis. https://www.clinicalstudydatarequest.com is a repository that hosts data from industry-funded clinical trials, where interested researchers can access data for more than 3,300 studies in a single place. In the field of cancer research specifically, where huge cost and effort are needed to do clinical trials, the Project Data Sphere platform created on 2014 to host data from more than 72 oncology trials and more than 150 data sets (Bertagnolli et al., 2017). The platform is available to almost any researcher with interest and experience in cancer research. Anyone interested in cancer research can apply and gain access to the data as well.

Moreover, several other journals now specialize in publishing data that can be reanalyzed by other researchers; these journals are called *data journals*. The following list presents the leading journals, with brief descriptions from their home pages:

1. *BMC Research Notes* (BioMed Central): An open-access, peer-reviewed journal that publishes short publications, cases, updates, software, databases, and data sets in the fields of biology and medicine.
2. *Dataset Papers in Science* (Hindawi): This journal published papers on science and medicine. This journal started in 2013 and has ceased publication in 2017 and is no longer accepting submissions, all of its previously published articles are archived.
3. *GigaScience* (BioMed Central): Publishes studies from the entire spectrum of life and biomedical sciences.
4. *Journal of Open Psychology Data (JOPD)* (Ubiquity Press): Features peer-reviewed papers describing psychology data sets with high reuse potential.
5. *Open Health Data* (Ubiquity Press): Publishes peer-reviewed papers describing health data sets with high reuse potential.

Implicitly available open data

Implicitly available open data need to be processed before reaching an analyzable form. The sources of these data can include social media websites (e.g., Twitter), service-providing websites (e.g., Google), and publication databases (e.g., Google Scholar). The steps required to achieve analyzable data vary depending on the targeted data source, it might be as simple as searching a database on Google Scholar, or as complicated as using specialized software to extract data from Twitter.

The sources of implicitly available open data are rapidly evolving, and their evolution depend mainly on the processing step required to reach the target data. The following are major groups of implicitly available open data; this list is by no mean conclusive, and readers should keep in mind that new entities keep being created. The idea from these examples is to describe this type of open data and to stimulate innovation for using it.

Social media content analysis – With recent technological advances and the spread of Internet access, the use of social media websites became widely available worldwide. These websites store huge amounts of data (known colloquially as *big data*) that can be related to medical, economical, or even political issues. From the researcher's point of view, these collected data can represent a valuable resource to use for research, and analyzing its content can generate a very large number of studies. For example, in a study that analyzed Twitter messages about influenza, the authors managed to forecast future influenza rates with an accuracy of 95% in relation to national health statistics (Culotta, 2010). The processing step used in this instance was extracting tweets containing flu-related keywords in order to collect a large number of Twitter messages in a short period of time through a combination of queries. The research method of studying documents and communications in various forms (i.e., text, audio, or video) is known as *content analysis* (Bryman and Bell, 2015).

For the health and biological sciences, social media are commonly used to obtain medical information (e.g., searching about certain diseases and treatments on weblogs like WebMd (www.webmd.com)), discuss health-related issues and expressing feelings about it (e.g., expressing pain during a disease using Facebook and Twitter), and even use tools to support clinical decisions such as UpToDate (www.uptodate.com). These social media websites store information regarding users' characteristics, frequency, and trends of usage (Denecke and Nejdl, 2009). Any researcher who wants to use these stored big data needs to put them in analyzable form, which can be done using an application designed for this purpose (either applications created by previous studies or commercially available ones such as Radian6 (www.radian6.com)) or software programs created via a programming language such as Python (Neuendorf, 2016).

Bibliometric analysis – Publications including scientific articles, books, conference papers, and other writings are usually stored in literature databases. Among these databases, the most commonly used and the largest ones are Google Scholar (www.scholar.google.com), Scopus (www.scopus.com), and Web of Science (www.webofknowledge.com) for almost all disciplines (Falagas et al., 2008), and PubMed (www.ncbi.nlm.nih.gov/pubmed) for biomedical and life sciences (Guz and Rushchitsky, 2009). The use of statistical methods to analyze these literature databases is known as *bibliometric analysis* (Young, 1983). An example of doing research using bibliometric analysis is analyzing research in a subject such as dry eye over a certain period of time (Boudry et al., 2018), analyzing research in general medicine publications (Kulkarni et al., 2009), or analyzing research by a particular country such as Jordan during a period of time (AlRyalat and Malkawi, 2018).

A scientist intending to do research on literature databases (bibliometric analysis) also needs to convert it to analyzable form. This step varies depending on the specific database; for instance, Google Scholar is a freely accessed, multidisciplinary database, but it provides less built-in analyzing functions than other databases (Jacso, 2005). Meanwhile, both Scopus and Web of Science are multidisciplinary, subscription-based databases, and they provide advanced analyzing tools (Falagas et al., 2008), and the PubMed database is free and provides some analyzing tools (Guz and Rushchitsky, 2009). A complete guide on using literature databases for bibliometric analysis with a video guide is published in the video journal *JoVe*.

Ideas from open data

This section will detail the steps for coming up with a research idea from open data (scenarios 1 and 2 in the previous section). As stated before, even if you do not have a specific idea, you might be interested in a certain specialty (e.g., neurology) or disease (e.g., cancer). So the first step is to choose the resource from the open data repository that you think best fits your interests.

Now, after you have chosen your data repository, you need to choose a particular subject (if you have not already done so). First, thoroughly review the subject in general, then look at published articles that have used the same data. We call this step a *double read*. For example, suppose that you want to do your research on sarcoidosis using the Biologic Specimen and Data Repository Information Coordinating Center (BioLINCC) open data repository. The first part of the double read is reading about sarcoidosis in general on resources such as Medscape (www.medscape.com) and UpToDate, and studying recent review articles to get a more in-depth understanding of the subject. An excellent source for this is the Cochrane Library (http://www.cochranelibrary.com), a source for high-quality, freely available articles that summarize what is already known and done in a particular subject of interest. You can find a list of such articles at the same page as your data of interest. You can also find publications that used the same data via search engines such as Google Scholar and PubMed, using keywords related to the data (e.g., "SPRINT" for the Systolic Blood Pressure Intervention Trial).

Most studies using open data have the name of the data as part of their title or abstract, or as a link usually provided in its acknowledgment section to the data repository website. The goal of the second part of the double read is gaining insight into what previous researchers did using the open data in the area you are interested in, and also to keep from replicating what others have already done.

At this point, you should have a general idea about your subject of interest, the open data you want to use, and what previous researchers published based on the open data that you have accessed. Your next step is the deep read, in which you decide what your research project will be about. During this step, after every study you examine, you should see if it has provided you with any new ideas. This step is similar to doing a conventional literature review as detailed later in this chapter, except that you must focus on studies' limitations and what they did not cover, in order to give you inspiration for your own research.

Here are strategies to help you to come up with a more specific idea:

1. During the second part of the double read, focus on the limitations of the current studies and suggestions for future studies. The limitations are often located just prior to the conclusion of a paper.
2. Read the variables available in your data set (e.g., age, gender, and history of heart attack) and try to think about studying an association between them.
3. Think outside the box. Try to link variables not related to the main objective of the original study. For example, in a study that was originally done to study cardiovascular disease, aim to find an association between body mass index (BMI) and level of education.

An example of the third strategy is a study that used data from the BioLINCC repository, specifically from the ACCESS data set (https://biolincc.nhlbi.nih.gov/studies/access/?q=access). In this study, authors used a data set to find the cause of sarcoidosis, but they studied the data provided to assess the relationship between chest x-ray findings in sarcoidosis patients and their demographics, an aim far from the goal of the original study (AlRyalat et al. 2017). An example of the first strategy also can be found in that study, where the authors stated the limitations and future goals as follows:

1. First, computed tomography scan can better evaluate abnormalities reported in this study that are difficult to characterize through X-ray, especially lung infiltrate.
2. Second, addressing clinical and laboratory associations with each imaging finding, especially the most commonly used scadding stages, will further increase the value of imaging in sarcoidosis.

These two points can be used by other researchers as a launching pad for a study that addresses them.

Keep in mind that even this more specific idea can be rather broad (e.g., the relationship between sarcoidosis and depression). But the important thing is that you need to read about the two topics (sarcoidosis and depression) thoroughly. You can always ask experts in the field to confirm the validity of your ideas, but remember that nothing can help you more than reading existing studies on your own.

It is important to keep your research focused—that is, include only one primary objective. The primary objective is the main idea your research is based on and the primary purpose for which you need the data. This primary objective will be translated into your research question (as discussed further in Chapter 2). Sometimes you might have secondary objectives that can be thought of as by-products of your research; they are not your primary focus, but you can find them during your research.

It is important to differentiate between the primary and secondary objectives during your research. One way to do this is by stating your primary objective, but not the secondary objectives, in the abstract, do not cover both the primary and secondary objectives during the results and discussion, and keep most of the discussion focused on the primary objective (Vetter and Mascha, 2017).

After determining an idea to base your research on, continue your literature review with a focus on discovering everything that has already been done in the subject area. Remember that you are going to write a paper that experts in the field may read to obtain more information, so you need to spend at least several days reading articles through a literature review to become thoroughly familiar with the topic yourself. These actions and the steps mentioned thus far in this section must be performed prior to actually requesting access to the open data.

Review (the deep read)

Conducting a quality literature review means that you need to *fit the literature into your research*. One of the best ways to conduct a quality literature review is the concept-centric approach (Webster and Watson, 2002). In such an approach, you base your review of the literature on the ideas in your research, unlike an author-centric or chronological-centric approach.

Figure 1.1 shows the three main steps in literature review. The first step is defining the keywords on which you will base your search of the literature; the proper choice of these keywords is essential to finding the research papers you want. You can choose the keywords based on the research question you formulated or its answer. Optimally, choose the three keywords that best fit your topic.

To further optimize choosing keywords, you can use a medical subject heading (MeSH) database (https://www.ncbi.nlm.nih.gov/mesh) to choose indexed terms, and there is an associated tutorial video to help you use the

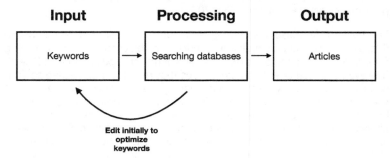

Figure 1.1   The three main steps in literature review begin with determining the keywords related to your research. These keywords will be used in the second step, which is searching databases. You can edit your keywords after an initial search to optimize the keywords according to your research. The third step is extracting articles related to your research from the search results.

site most effectively. Since different authors use different terminologies to refer to the same concept (e.g., "cancer" and "tumor" are sometimes used interchangeably in papers), PubMed uses a standard vocabulary system for retrieval of information. The most efficient way to search PubMed is by using MeSH terms in the query. An example of the benefits of using MeSH terms instead of basic searching is when you want to look for studies in Jordan, choosing "MeSH terms" to search specifically for the keywords you want will not retrieve articles authored by someone named "Jordan" when you are actually searching for articles published in the country of Jordan. Remember that the better your input keywords are, the better the output articles are, and vice versa (a concept known as "garbage-in/ garbage-out").

The second step is executing a literature search on the related databases, most commonly Google Scholar or PubMed. During the search, screen the titles for their relevance, screen the abstracts of the articles with relevant titles, and then read all articles that have relevant titles and abstracts. Finally, collect the useful information from each article.

To decide which information is important to your study, continuously ask yourself the following question:

"How is the work presented in the article I read related to my study?" Answering this question will allow researchers to fit the literature into their own research (Levy and Ellis, 2006). Once you conclude that you are getting the same findings over and over again from the articles you are reading, you should finish the literature review and proceed to the next step (Webster and Watson, 2002).

To further stress the importance of quality literature review, recall the famous statement of Sir Isaac Newton (1676): "If I have seen a little further, it is by standing on the shoulders of giants." Obtaining high-quality articles from your literature review will let you expand your knowledge on a subject, unlike poor-quality articles (Levy and Ellis, 2006).

The leading database search engine for biomedical literature is PubMed (https://www.ncbi.nlm.nih.gov/pubmed/). It can be used by simply searching for keywords using the basic search screen at its home page (Figure 1.2, left); however, the results will contain articles that might not be related to your work (i.e., the results aren't specific enough). Alternatively, the advanced search builder (Figure 1.2, right) will save time by finding the articles that best suit your needs (Fatehi et al., 2014). The advanced search builder allows a targeted search in specific fields, with the convenience of being able to select the intended search field from a list. In the "Advanced search form," you can add terms related to your search and use three operators to determine the relation between the entered terms:

- AND: Indicates that both terms around it must appear.
- OR: Indicates that at least one of the terms around it must appear.
- NOT: Indicates that the term after it must not appear.

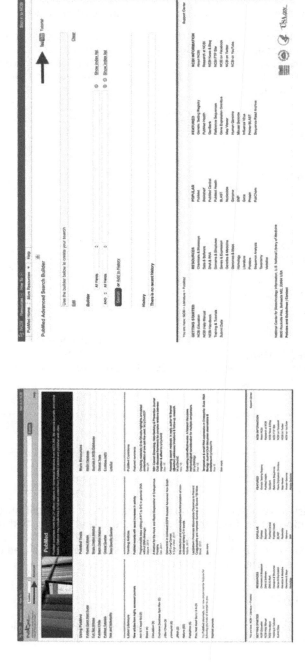

Figure 1.2 Doing an advanced search on PubMed will filter studies to find the ones most related to your subject. Left: The home page of PubMed, where you can do a simple search. Clicking "Advanced search" (arrow) will take you to the Advanced Search screen. Right: The Advanced Search screen allows you to type the keywords related to your topic, and a video tutorial at the top right provides simple tips.

After clicking "Search" and getting the results, you can further limit the results according to several filters (e.g., publication in a custom period).

The final step of literature review is obtaining the output. It is advised to fill in a form with the results of the included studies, so you can use them as you write your manuscript. It is also advised to keep all articles in a folder, for easy access. The following form is suggested initially until you make your own personalized version:

Literature Review

Title of the study:

Date of PubMed search:

Keywords used:

Search strategy:

Link of the PubMed search:

# PubMed	First Author, Year	Study Aim	Conclusion	Findings Related to Our Study	Notes

Last but not least, it is worth having a brief discussion about the importance of extracting information from the original source. That means that if information you read in an article actually came from another article, you should go to that source and cite that. Eventually, your citation is a credit, so you should give this credit to the original study.

Access open data

As already mentioned, the open data model was suggested to allow maximum benefits from research. However, the main disadvantage of the open data model is its potential misuse by nonprofessionals. For this reason, the gatekeeper model came into being (Rockhold et al., 2016). This model stipulates that distinct entities keep information in central repositories (e.g., websites), with access to specific data sets provided only to qualified research teams on the basis of a review of their research proposals by an independent expert committee. This model can assure the appropriate use of data by professionals in the field, and the prospect of a study being the source of data citation can serve as an incentive for the original investigators.

Repositories vary in the application of this gatekeeper model. Some repositories, usually public repositories like the Dryad data repository (www.datadryad.org), allow anyone to download data directly from the repository without any documentation needed and without restrictions. Some repositories, on the other hand, require several documents, including your résumé, your study's protocol (as detailed in Chapter 2), and your study's ethical approval (detailed in Chapter 2), so that only qualified researchers can access these data sets. These repositories are usually institution-affiliated repositories such as YODA.

Almost all open data repositories have a manual file on how to use the data properly. One example is the BioLINCC repository handbook found on the BioLINCC repository website (https://biolincc.nhlbi.nih.gov/home/). Read it thoroughly, as well as the primary publication associated with the data that you chose. The data will be available in several forms (e.g., Excel, Statistical Package for the Social Sciences (SPSS), or Statistical Analysis System (SAS)), but you won't access these forms unless you access the data, so you will need to depend on the associated manuals and publications to learn what variables and measures are included in the data set.

After having a more specific idea, you want to get access to the open data that you chose. First, check the requirements and the steps for accessing the data from the repository. You need to prepare yourself for three main requirements for any repository:

1. Your information and qualifications (i.e., résumé or CV)
2. The idea (in the form of a protocol, as detailed in Chapter 2)
3. Institutional approval, such as from an institutional review board (IRB)

Always keep the idea in your protocol simple; you should not exaggerate what you are planning to do just to gain data access. Open data repositories are created for the purpose of providing access; they just need to make sure that you will use the data properly. For almost all NIH-associated repositories, an electronic application form must be filled out and the abovementioned documents uploaded. Each application is carefully reviewed to ensure the investigators' affiliation with scientific or educational institutions, and to ensure that the proposed research idea for data use has not already been done. Incomplete applications will not be approved. The results of the committee's review will be sent to you via email. Approved applicants will receive login information to access and download the data. The review process varies, but takes an average of 1–2 weeks, during which you should continue reading and saving relevant articles, as you will come back to them later in your study.

Ethical considerations on open data

As for any other research, a study that is done using open-access data should adhere to certain regulations designed to protect participants' data and to ensure that the data will be used properly and safely. Up to this moment, there are no universal regulations to protect open-data research worldwide; however, each open-data repository has its own regulations with which each applicant/researcher must comply.

Any study participant expects that certain measures will be implemented to protect his or her data, including data anonymization and informed consent to explain how the data will be used and managed. On the other hand, some kinds of data are peculiar and these measures may not be enough; genomic data represent the most important example of this. Moreover, primary researchers who collected and managed the data expect an acknowledgment from the secondary users of these data sets as well. In this section, we will summarize most of the relevant regulations governing open data.

Data anonymization

One of the main concerns about dealing with open data is that personal information might be disclosed without permission. This issue was addressed by anonymizing the data; that is, requiring the removal of individual identifiers to an extent that prevent any identification of an individual and preventing researchers accessing the data from trying to deanonymize the data (U.K. Government, 2012). Several institutions have stressed the importance of data anonymization. For example, the University of Sheffield's policy on reusing existing data states that all researchers are strongly encouraged to keep in mind the possibility of secondary research and data sharing at the outset, before primary data collection begins, and to build this concept into the informed consent process (https://orda.shef. ac.uk). The university further states that "data can be shared as long as it is completely anonymous such that individual respondents cannot be identified."

So long as data are appropriately anonymized to protect confidentiality, they should be freely available to download and analyze the data for research purposes. This practical approach to data sharing is known as the *open source model* (Bertagnolli et al., 2017). Moreover, some data repositories provide data access under Creative Commons 0 (CC-0; https:// creativecommons.org/publicdomain/zero/1.0/), which states, "The person who associated this particular work with this deed has dedicated the work to the public repository by waiving all of his or her rights to the work worldwide under copyright law, including all related and neighboring rights, to the extent allowed by law." This is in contrast to the gatekeeper model, which restricts data access to qualified researchers and provides data owners control over their data. On the other hand, an overwhelming number of application forms and preliminary documents are sometimes required before data access is granted.

In the repository holding the most data sets on clinical trials (ClinicalStudyDataRequest.com), the most common cause for some sponsors to not being able to provide a study's data is "High risk of de-identification," which might occur for many reasons, including the small number of patients included in the trial, poor anonymization, or the use of demographic data for a famous clinical trial. The common goal of all data-sharing models and policies is to protect patients' personal information by keeping all data anonymized, so researchers should never try to contact or deanonymize any study participants.

Informed consent and confidentiality of data

Through informed consent, patients acknowledge that the data collected in the primary study in which they are involved will be shared and may be used by other researchers through secondary research. Participants in the primary study are assured that their data will be anonymized and secured and won't be used for other than research purposes.

Most repositories ask their applicants to take certain measures to ensure the security of data, such as storing data on a secured computer, to only allow approved and supervised persons to have access to the data, and to discard all data documents and biological specimens after a certain period of time. Data repository agencies also require researchers who are seeking access to data to fill out agreements and application forms, to agree to a review of their study and its investigators and protocol, and to adhere to the agencies' regulations and provisions. Most repositories also do not provide access to an entire data set; researchers must specify what parts of the data they need to perform their study.

Acknowledgment of data owners

In the recent past, a data set would have been used primarily by the researchers who had actually created it, and it would provide the basis for these researchers to create many publications. There would have been a direct relationship between data creation and control over its usage and the publication of its results. However, with the data-sharing policies implemented at this time, the fact that particular researchers have created a data set no longer gives them enduring control over its use and resulting publications. The challenge, then, is how to reward and acknowledge the production and sharing of a data set (Kaye et al., 2009).

The traditional way of acknowledging a research source is through citation. Usually, upon obtaining open data, you will have to agree to acknowledge the contribution of the data to all oral and written presentations, disclosures, or publications resulting from any analyses conducted on the data. This acknowledgment usually should be done in an explicit way, with a wording similar to the following:

> This study was prepared using --------- open data obtained from the ---------- data repository.

Genomic data

The field of genomics is unique with regard to data sharing and open research, as it requires large numbers of samples, comparative populations, and advanced infrastructure and computing technologies. These and other characteristics made the field of genomics the leader in data-sharing policy and promotion (Kaye et al., 2009). An example of an international collaboration with a data-sharing strategy is the Human Genome Project (HGP) launched in 1990, an international, collaborative research program whose goal was the complete mapping and understanding of all the genes of human beings (NIH, 2005). Every part of the genome sequenced by the HGP was made public immediately, and new data on the genome is posted every 24 hours. Moreover, funders of genome-related projects require that data sharing be considered for every project unless there are justifiable reasons otherwise. Examples of such funders with these policies are the NIH, Genome Canada, and the Medical Research Council (MRC) in the United Kingdom. All of these organizations now make data sharing a requirement of funding for genomics research projects. With these policies in mind, the question for researchers has become how to share data, whereas previously it was whether they should be shared at all (Kaye et al., 2009).

The first international document to lay out the principles for open access in the field of genomics was the 1996 Bermuda Agreement. The key idea being promoted in this agreement is that the prepublication genome sequence should be freely available and in the public domain in order to encourage research and development and to maximize the benefit to society (Kaye et al., 2009). The traditional mechanisms that have been used to protect research participants' data, informed consent and data anonymization, may not be enough to protect genomic data. An example of such difficulty is a study that showed the possibility of identifying a specific person when less than 0.1% of the total genomic data is identified, even if his or her data are demographically anonymized, and that it is less difficult than it should be to assess the probability that a person or relative participated in a genomic study (Homer et al., 2008). These findings led both the NIH and the Wellcome Trust in the United Kingdom to remove some genomic data from publically available repositories (*Nature* Editors, 2008). Regarding data anonymization for genomic studies, procedures for controlling disclosure (e.g., coding each study subject or aggregating the information) can be further employed to protect the identity of data subjects (Kaye et al., 2009).

The following projects are among the largest open access genomic databases:

- Human Genome Project (HGP): https://www.genome.gov/12011238/an-overview-of-the-human-genome-project/
- The HapMap Project: http://www.hapmap.org/abouthapmap.html
- The 1000 Genomes Project: http://www.1000genomes.org/page.php

Tips on open data research

The following points are quick tips that should be kept in mind when doing open data research:

- Open data saves you time on data collection. Invest this time wisely, to come up with a novel idea—one that is not an obvious extension of the reported work.
- Some data repositories (e.g., NIH-associated repositories) prohibit mentioning the name of the original data set in the title of your paper, and whereas some guidelines allow the name to be used in the title or the abstract, try always to put it only in the abstract.
- Researchers are usually required to report immediately and in detail any change in their study plan, any unanticipated problems, or any change in the investigators list that was previously disclosed to the repository upon submission for access.
- Send a copy, either to the repository website or via email to the data set's owner, of the published study using open data as a courtesy.

Data tutorial

In this tutorial, we will follow the steps discussed in the previous section to generate ideas from a data set. This demonstration is for teaching purposes only.

Let us begin by choosing the data that best fits our interest. Suppose that we would like to do a study related to heart attacks, as this is a general topic of interest to many people in many specialties. First, we need to look for a high-quality data set related to this subject by looking in the most appropriate data repository. As we stated in this chapter, of the data repositories available, NIH-associated ones include the highest-quality data sets for research funded by billions of dollars annually. By looking on the BMIC website, which stores all data repositories associated with the NIH, we can see that the NHLBI repository BioLINCC is the repository most suitable for us. This repository provides the option of searching data sets based on a disease category, where we can choose "Cardiovascular diseases" as the category of interest. From the search results, one of the data sets included is from the Framingham Heart Study. We will base our study on this data set.

After choosing the best data set for our subject, now we need to do the double read. First, read about heart attacks in general to get a sense of your subject. We should begin by searching sites such as Medscape and UpToDate, and then search for recent review articles on sites such as the Cochrane Library. Next, we should read other published articles that have used data from the Framingham Heart Study. Most articles that use open data have the name of the original study as part of their title or abstract, so we can search Google Scholar or PubMed using the term "Framingham Study." The more studies we read, the more insight we gain into our subject.

While doing the double read, we need to keep in mind potential research ideas. In the previous section, we discussed three main strategies you can use to come up with your own idea.

An example of how to use the first strategy involves a study about the predictive accuracy of the Framingham Coronary Risk Score in British men (Brindle, 2003). In this study, the authors stated in the "Discussion" section that one of their limitations was that they could not generalize their findings to women. We could take advantage of this limitation by doing our own analysis on the female population in the Framingham Heart Study.

By exploring the Framingham Heart Study website, we can use the second strategy by reading about the general demographics, BMI, and history of cardiovascular events. At this point, we could come up with an idea to study the difference between the BMI of patients with a previous history of cardiovascular events and the BMI of patients without such a history.

The third strategy stems from methods used in the second strategy. However, the difference is that you are assessing the association between variables not related to the main objective of the original study. For example, you might assess the association between a patient's BMI and level of education.

For our study, let us choose the idea from the second strategy.

Now we need to gain access to the Framingham Heart Study data. The Framingham Heart Study is part of the BioLINCC data repository. For NIH repositories, you will need to fill out an online application form that requires four main documents:

1. The research protocol, which should detail how we are going to use the data for the research idea we have (it will be the subject of the tutorial in Chapter 2)
2. Your résumé/CV, to document that you are qualified to use the data to do the proposed study
3. An ethical approval, which can be obtained from your institution (more details are provided in Chapter 2)
4. An agreement between you and the NIH representatives for the optimum use of data

At the time of writing, it usually takes 7–10 days for the independent expert committee to review and approve of an application to access the data. That includes reviewing the qualification of the researcher, appropriateness of the data for the proposed research, and compliance with human subject regulations. After that, you will gain access to the data, which should be in a machine-readable form such as an Excel spreadsheet, where you can analyze the data (as detailed in Chapter 3), and finally write and publish the manuscript (as detailed in Chapter 4).

References

AlRyalat SA, Al-Essa M, Ghazal R, Abusalim E, Mobaideen D, Alrahmeh S, Alatrash M, Obaidat N. Chest X-ray in sarcoidosis: The association of age, gender, and ethnicity with different radiological findings. *Current Respiratory Medicine Reviews*. 2017 Dec 1;13(4):241–246.

AlRyalat SA, Malkawi L. International collaboration and openness in Jordanian research output: A 10-year publications feedback. *Publishing Research Quarterly*. 2018 Jun 1;34(2):265–274.

Ambrosius WT, Sink KM, Foy CG, Berlowitz DR, Cheung AK, Cushman WC et al. The design and rationale of a multicenter clinical trial comparing two strategies for control of systolic blood pressure: The Systolic Blood Pressure Intervention Trial (SPRINT). *Clinical Trials*. 2014 Oct;11(5):532–546.

Bertagnolli MM, Sartor O, Chabner BA, Rothenberg ML, Khozin S, Hugh-Jones C et al. Advantages of a truly open-access data-sharing model. *New England Journal of Medicine*. 2017;376:1178–1181.

Boudry C, Baudouin C, Mouriaux F. International publication trends in dry eye disease research: A bibliometric analysis. *The Ocular Surface*. 2018 Jan 1;16(1):173–179.

Brindle P, Jonathan E, Lampe F, Walker M, Whincup P, Fahey T, Ebrahim S. Predictive accuracy of the Framingham coronary risk score in British men: Prospective cohort study. *Bmj*. 2003 Nov 27;327(7426):1267.

Bryman A, Bell E. *Business research methods*. New York: Oxford University Press, 2015.

Burns NS, Miller PW. Learning what we didn't know—The SPRINT data analysis challenge. *New England Journal of Medicine*. 2017 Jun 8;376(23):2205–2207.

Byrd J. Data sharing models. *New England Journal of Medicine*. 2017;376:2305–2306.

Culotta A. Towards detecting influenza epidemics by analyzing Twitter messages. In *Proceedings of the First Workshop on Social Media Analytics*, 2010;115–122. ACM New York, New York.

Dawber TR, Meadors GF, Moore FE Jr. Epidemiological approaches to heart disease: The Framingham Study. *American Journal of Public Health and the Nation's Health*. 1951 Mar;41(3):279–286.

Denecke K, Nejdl W. How valuable is medical social media data? Content analysis of the medical web. *Information Sciences*. 2009 May 30;179(12):1870–1880.

Dicks LV, Walsh JC, Sutherland WJ. Organising evidence for environmental management decisions: A '4S' hierarchy. *Trends in Ecology & Evolution*. 2014 Nov 1;29(11):607–613.

Dietrich D, Gray J, McNamara T, Poikola A, Pollock P, Tait J, Zijlstra T. *Open data handbook*. Open Knowledge International; 2009.

Editors. DNA databases shut after identities compromised. *Nature*. 2008;455:13.

Falagas ME, Pitsouni EI, Malietzis GA, Pappas G. Comparison of PubMed, Scopus, Web of Science, and Google Scholar: Strengths and weaknesses. *The FASEB Journal*. 2008 Feb;22(2):338–342.

Fatehi F, Gray LC, Wootton R. How to improve your PubMed/MEDLINE searches: 3. Advanced searching, MeSH and My NCBI. *Journal of Telemedicine and Telecare*. 2014 Mar;20(2):102–112.

Guz AN, Rushchitsky JJ. Scopus: A system for the evaluation of scientific journals. *International Applied Mechanics*. 2009 Apr 1;45(4):351.

Homer N, Szelinger S, Redman M, Duggan D, Tembe W, Muehling J et al. Resolving individuals contributing trace amounts of DNA to highly complex mixtures using high-density SNP genotyping microarrays. *PLoS Genetics*. 2008 Aug 29;4(8):e1000167.

Horton R. North and South: Bridging the information gap. *The Lancet*. 2000 Jun 24;355(9222):2231–6.

ISSC, IDS, and UNESCO. *World social science report 2016*. Challenging inequalities: Pathways to a just world, 2016.

Jacso P. As we may search—Comparison of major features of the Web of Science, Scopus, and Google Scholar citation-based and citation-enhanced databases. *Current Science*. 2005 Nov 10;89(9):1537–1547.

Kaye J, Heeney C, Hawkins N, De Vries J, Boddington P. Data sharing in genomics—Re-shaping scientific practice. *Nature Reviews Genetics*. 2009 May;10(5):331.

Kulkarni AV, Aziz B, Shams I, Busse JW. Comparisons of citations in Web of Science, Scopus, and Google Scholar for articles published in general medical journals. *JAMA*. 2009 Sep 9;302(10):1092–1096.

Levy Y, Ellis TJ. A systems approach to conduct an effective literature review in support of information systems research. *Informing Science*. 2006 Jan 1;9.

Li D, Azoulay P, Sampat BN. The applied value of public investments in biomedical research. *Science*. 2017 Apr 7;356(6333):78–81.

Manyika J, Chui M, Brown B, Bughin J, Dobbs R, Roxburgh C, Byers AH. Big data: The next frontier for innovation, competition, and productivity. 2011 McKinsey & Company. https://www.mckinsey.com/~/media/McKinsey/Business%20Functions/McKinsey%20Digital/Our%20Insights/Big%20data%20The%20next%20frontier%20for%20innovation/MGI_big_data_exec_summary.ashx.

Martin L, Hutchens M, Hawkins C, Radnov A. How much do clinical trials cost? *Nature Reviews. Drug Discovery*. 2017;16:381–382.

National Heart, Lung, and Blood Institute (NHLBI). SPRINT overview. November 2017, https://www.nhlbi.nih.gov/news/systolic-blood-pressure-intervention-trial-sprint-overview

National Human Genome Research Institute. An overview of the Human Genome Project. October, 2005. Accessed May 5, 2006, from http://www.genome.gov/12011238 Washington, DC.

Nelson B. Data sharing: Empty archives. *Nature News*. 2009 Sep 9;461(7261):160–163.

Neuendorf KA. *The content analysis guidebook*. Sage; 2016.

Piwowar HA, Day RS, Fridsma DB. Sharing detailed research data is associated with increased citation rate. *PLoS One*. 2007 Mar 21;2(3):e308.

Rockhold F, Nisen P, Freeman A. Data sharing at a crossroads. *New England Journal of Medicine*. 2016 Sep 22;375(12):1115–1117.

U.K. Government. *Open data white paper: Unleashing the potential*, https://www.gov.uk/government/publications/open-data-white-paper-unleashing-the-potential, 2012.

Vernon MM, Balas EA, Momani S. Are university rankings useful to improve research? A systematic review. *PloS One*. 2018 Mar 7;13(3):e0193762.

Vetter TR, Mascha EJ. Defining the primary outcomes and justifying secondary outcomes of a study: Usually, the fewer, the better. *Anesthesia & Analgesia*. 2017 Aug 1;125(2):678–681.

Webster J, Watson RT. Analyzing the past to prepare for the future: Writing a literature review. *MIS quarterly*. 2002 Jun 1;xiii–xiii.

World Bank. Research and Development Expenditure (% of GDP). Accessed in May 2018 from https://data.worldbank.org/indicator/GB.XPD.RSDV.GD.ZS?locations=JO-XM&year_high_desc=false

Young H. *The ALA glossary of library and information science*. Ediciones Díaz de Santos; 1983.

2

Essential Research Background

Saif Aldeen Saleh AlRyalat and Lna W. Malkawi

Research overview

Research is defined as an endeavor that scholars intentionally perform to enhance their understanding of a phenomenon, and they expect to communicate what they discover to the larger scientific community (Leedy and Ormrod, 2005). It also can be defined as "[a] process of steps used to collect and analyze information to increase our understanding of a topic or issue" (Creswell, 2008). In other words, research consists of answering a question that you have via collecting and analyzing data to generate results. Research is usually communicated through reporting the results in the form of a manuscript that can be published in a public forum, such as an article in a journal. The following real-life case scenario will further clarify the idea of doing research (Alhawari et al., 2018).

A researcher at the University of Jordan observed a high frequency of overweight and obese students at the school, and he already knew that high body weight leads to hypertension (high blood pressure). He asked the following question "What is the frequency of hypertension among university students?" After learning how to do research, he conducted the steps of collecting and analyzing data (which will be explained in the upcoming chapters of this book) to find that there was indeed a high frequency of hypertension among students at the university. To communicate these results so other researchers and policymakers can benefit from them, he put together a manuscript of a report, which he then sent to a journal focusing on the topic of hypertension. The journal reviewed his manuscript and decided to publish it. Now, his research has culminated in an article in a journal, which anyone interested can read to benefit from his findings.

This example summarized the journey of doing research, from the first step (i.e., an observation), to the final step, when the observation become an article published in a journal. The best-case scenario is that the researcher in the example finds a good mentor to guide him through his research journey, but that does not always happen. The main purpose of this book is to give you the information you will need to pursue your research journey with minimal mentoring. Your ability to do so will depend on your interest and inclination to learn by following the examples presented here and reading more on the topic of research in other sources. The more you read, the more you will learn, especially if you have a real desire and motivation to do so.

Choose the topic you are most interested in and read published articles on it before beginning your own research. One of the best ways to do this is to use research-related social media, including ResearchGate (researchgate. net). One of the advantages of these websites is that researchers add full articles on it, and access is free of charge, so all you need to do is choose your topic of interest and read the related articles. When doing so, you will want to pay attention to the "Methodology" section, where authors describe how they got their results, and to evaluate the quality of the journal an article is published in (for instance, if it has a good reputation).

These days, doing research and publishing reports on it are part of the criteria that determine the success of an academic institution. As a result, these institutions mandate frequent publishing from its staff in order to promote its teaching staff and graduate its students. Moreover, frequent publication becomes one of the few powerful ways that scholars can demonstrate their talents and knowledge to their peers (Rawat and Meena, 2014). The phrase *publish or perish* was coined to describe the continuous pressure on academic staff to publish scholarly research, even without having the resources, support, or even the knowledge required to do so.

At this point, it is good to keep in mind a basic principle; Publishing in a good (i.e., high-impact) journal requires high novelty results built on strong methodology, so try your best from the beginning to get novel results and design a strong methodology. If you want your article to have a good reputation, you will need to publish in a good journal (Figure 2.1); Chapter 4 will describe in detail how to choose a journal. For instance, if the name of a journal is listed on the Scopus search engine (Scopus.com), then it is a fairly a good journal.

But why would anyone need to understand or do research?

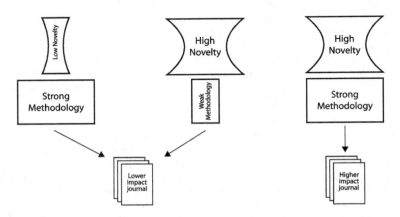

Figure 2.1   High-novelty results that are based on a strong methodology will yield an article published in a higher-impact journal compared to low-novelty results and a weaker methodology.

Research is important for your career as a scientist, for your institution, and for your country. The benefits of doing research for you are the following:

- Doing and understanding research will give a sense of what you are studying and/or teaching, as you will know the basis and the journey that lie behind any information you obtain.
- Research will also teach you how to judge the validity of any information you encounter, rather than just taking this information for granted.
- Most of the time, research is mandatory for graduation and promotion.
- The more you do research, the higher chance you will have to get scholarships and better positions.
- Finally, being a researcher can open the door into an industry career, mainly in the research and development team in a pharmaceutical company (Melese et al., 2009).

The benefits of doing research for your institution are the following:

- The strength of an institution (e.g., a university) is measured mostly through the number of published studies it produces (Vernon et al., 2018).
- The more research an institution produces, the greater the chance that it can get grants and funds to support its efforts. From 1994 to 2003, the expenditure on biomedical research increased by around $70 billion per year, and the numbers have increased significantly since then (Moses et al., 2005).

The benefits of research for your country are that several indices measure the educational strength of a country and the strength of its certificates by the research production in this country (AlRyalat and Malkawi, 2018).

Before you start doing research, it's important to clear up several misconceptions about it:

- It is preferable to have a mentor during your first research, but it is not mandatory. You can read thoroughly about how to do research on your own, including consulting sources like this book.
- You can publish your first study in a good journal with high impact— don't think the chances of success are so low that you shouldn't even try.
- You do not need to spend a lot of time on collecting data; you can easily access data from outside sources that are ready to use for your research, on any topic.
- Do not be emotional as you write your manuscript, and always choose the shortest sentence that delivers your point.
- Writers are made, not born, unlike the misconception that "writers are born writers" (Troyka, 1993).

Study components

Any study, whether done with original research or open data, should have two main phases: planning (when you come up with your research question and develop a protocol to execute it) and execution (when you follow the protocol to answer your research question). Figure 2.2 summarizes the contents of this section.

The planning phase in open data research is very important, as you will need to use a protocol to access the data and obtain the necessary ethical approvals. It is important to prepare a high-quality protocol that describes your study and why you need the data. This chapter's tutorial will concentrate on preparing a protocol to access open data.

Research question

Scientific research must begin with a defined, open-ended question, so that you can respond to the question, "What is your research about?" by saying, "My research will answer the following question…" The question should meet three major requirements: operability, measurability, and having a clear description of your sample (Röhrig et al., 2009). For a question to be operable, you should have a general idea about how you are going to execute your research (via a questionnaire, for example). For a question to be measurable, you should have an idea about what information you will find after executing your research (e.g., age, gender, and pain score). Finally, upon formulating your question you should describe your target population (e.g., elderly patients) and any other specific conditions for your

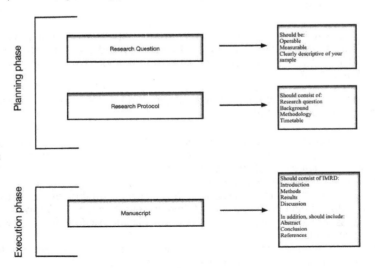

Figure 2.2   Summary of the structure of a study, depicting the two main phases.

research (e.g., talking to patients after surgery). An example of a research question that meets all the requirements is the following:

- Is there any difference in pain before and after the use of analgesia in elderly patients with dementia?
  - *Operability*: We will assess pain by distributing questionnaires at geriatric care centers.
  - *Measurability*: We will measure pain scores via questionnaires before and after the administration of analgesia.
  - *Description of the target population*: We will study elderly patients with dementia.

By default, your research question should define your primary objective (i.e., the main research aim). It is preferable to have one primary objective in your research so that your efforts will be focused (Vetter and Mascha, 2017).

Research protocol

After formulating a good research question, you now need to create a research protocol. A *research protocol* (a term sometimes used interchangeably with *research proposal*) is a document that explicitly states the underlying reasoning and structure of a research project (Moeen, 2008). Your protocol is a valuable practical timetable and guide to your research plan that gives insight into what you are trying to achieve. Above all, it requires advanced planning, anticipation of potential problems, and a plan to deal with them.

As illustrated in Figure 2.2, a protocol should have the following mandatory sections:

- A research question
- Background information based on a literature review of previous studies (as detailed in Chapter 1)
- Methodology that details what data you will use and how you will analyze these data to answer your research question
- The timetable that you intend to follow as you execute your research

In the classical gatekeeping model adopted by many open data repositories, including those associated with the National Institutes of Health (NIH), you must have a protocol for your study in order to gain access to the data, and the protocol will be used to evaluate the novelty of your idea and your ability to explore it. You also will use it to obtain ethical approval for your study from your institution. Sometimes repositories and institutions will have a special template for you to fill out to create your protocol, which should include the abovementioned sections. If no template is required, you can refer to the protocol written in the tutorial at the end of this chapter.

Manuscript

Up to this point, you have been in the planning phase of your research project. Only after completing this phase are you eligible to actually begin your research study (i.e., the execution phase). A manuscript should consist of four basic sections: Introduction, Methods, Results, and Discussion (IMRD). Other sections may be derived from the basic IMRD sections as well, including the following: an abstract at the beginning, which summarizes the IMRD sections in one paragraph; a conclusion which briefly states the major results of the research and their contribution to the overall field of study; and a list of all the references used in the study (Fisher et al., 2013).

A general approach to writing a scientific paper (the process of which will be detailed in Chapter 4) is to begin with writing the "Methodology" section, derived from your initial research protocol, detailing all the steps that you performed to reach your results. Next, write the "Results" section, which contains the data found from your work. Then, write the Introduction, and use the Introduction and Results sections to guide the writing of the Discussion section. Finally, summarize the contents of the paper in the Abstract, and then condense and refocus the Abstract according to the Conclusion section. As you create your report, remember that the scientific aspects of the manuscript may be very challenging for the reader to comprehend; therefore, you should work to convey the scientific information in a clear, understandable manner.

After completing these sections, the final task is to create a title page that is derived from and clearly describes your study and also includes the list of authors. The authors must be listed based on their contribution to and role in the study (lead author, corresponding author, etc.). Others who did not contribute significantly to the work but have provided some support (e.g., statistical analysis) should be mentioned in an "Acknowledgments" section. For more information about authorship, please see section 2.5.4, later in this chapter.

Types of studies

Knowing the type of study that you are reading will help you determine the level of evidence you are dealing with, to what degree you may depend on this study in your clinical practice, and what conclusions you may draw from it. The phrase *evidence-based medicine (EBM)* is used by physicians, researchers, policymakers, and others to describe the quality of the conclusion based on the design of the study that produced it (McNair and Lewis, 2012). Deriving evidence from research is done by two main authorities: the U.S. Preventive Services Task Force (USPSTF) and the Centre for Evidence-based Medicine (CEBM) at the University of Oxford. These authorities assign a specific level of evidence based on the study design and the quality of that study in a particular subject (Table 2.1).

For researchers, it is important to know the study design based on your research question. If, from the beginning, you do not choose the best study design according to the aim of your study, nothing can correct this mistake; it would be like building a palace on sand.

Table 2.1 Main Study Designs with Their Respective Relation to Time, Presence or Absence of an Intervention, and the Level of Evidence Derived from Them

Study Design	Perspective in Time	Experimental or Observational	Level of Evidence[a]	Brief Description
Randomized controlled trial	Prospective	Experimental	1	Randomly assign patients to either treatment or placebo
Nonrandomized (quasi-experimental) trial	Prospective	Experimental	2	Nonrandomly assign patients to either treatment or placebo
Cohort study	Prospective or retrospective	Observational	2	Observe the outcome in a group of patients with specific risk factors
Case-control study	Retrospective	Observational	3	Compare risk factors in patients with a condition and in patients without
Cross-sectional study	One point in time	Observational	4	Assess an outcome at one point in time
Case report	Retrospective	Observational	4	Detailed data of a case or cases

[a] Centre for Evidence-based Medicine—Levels of Evidence (March 2009).

Note: Remember that evidence from poor-quality studies are downgraded to a higher number.

The classification of study types depends on the perspective from which you are looking. In prospective studies, you are taking variables into the future, while in retrospective studies, you look at variables in the past (Dykacz, 2014). As illustrated in Table 2.1, we also may classify studies based on the presence or absence of an intervention (McNair and Lewis, 2012). In observational studies (e.g., cohort, case-control, cross-sectional, and case reports), you just observe, without performing any intervention, whereas in experimental studies (e.g., randomized controlled and nonrandomized trials), you are performing an experiment.

Finally, there are two other, completely different types of research: systematic review and meta-analysis (Ressing et al., 2009). A *systematic review* is a critical assessment and evaluation of all research studies that address a particular issue (e.g., lung cancer). On the other hand, a meta-analysis is a statistical examination combining information from multiple scientific studies and reanalyzing it to generate results from collective data. In both of these, researchers use an organized method of locating, assembling, and evaluating a body of literature on a particular topic using a set of specific criteria. These two types of studies may be combined: researchers can statistically analyze data from different studies (meta-analysis), and then do a systematic review of the body of literature.

As an early researcher, you will most likely be concerned more with conducting observational studies than experimental ones. Observational studies may have a massive impact that cannot be equaled by experimental studies. A great example of this occurred in 1936, when John Snow showed that cholera could be stopped by cutting off the water supply from contaminated wells. This was only discovered by observing the times and places where the disease occurred and the sources of the water supply to those areas. These observations were sufficient to pinpoint the major environmental factor inducing cholera. The preventive action of cutting off the contaminated water supply by removing the handles of certain specific water pumps put a stop to the disease.

Observational studies

There are four main observational study designs: prospective cohort, retrospective cohort, case-control, and cross-sectional. In this subsection, we will provide a general idea about each of these and their pros and cons, and then we will give guidelines on how to choose the best design for your study. We will present them in descending order according to the quality of data they provide (Omair, 2015).

Prospective cohort study – This requires you to follow a population with a specific exposure to determine the subsequent occurrence of a certain health outcome. The best example for this type of study is the Framingham Heart Study (Dawber et al., 1951), in which patients who were predisposed to developing cardiovascular disease (e.g., patients with hypertension) were followed for the development of outcomes such as stroke. Prospective cohort studies begin by recruiting patients based on preset inclusion criteria (i.e., the exposure you are studying, such as hypertension), and then following this group (called a *cohort*) for the predicted outcome (e.g., stroke). The advantages of this design include the fact that you can choose the exposure you need without restrictions, you might establish causal association in your study, and you can find the incidence of a disease. Hhowever, there are major feasibility constraints that come with following up with these patients, and thus this type of study needs significant time and resources compared to other study designs (Checkoway et al., 2007).

Retrospective cohort study – This is a practical alternative to the prospective cohort, performed by examining the records of a group of patients with certain exposures (also called a *cohort*) and checking for the predicted outcome (which also can be found in the records). An example for this approach is a study examining the occurrence of chronic kidney disease (outcome) in patients with renal tumors (exposure), in which authors retrospectively analyzed patients' data from hospital records (Huang et al., 2006). A good point for this design is its high feasibility, but an obvious problem is the potential absence of data on the patients' past exposures to disease (Checkoway et al., 2007).

Case-control study – This type of study may be thought of as the opposite of cohort studies. Here, you are reviewing patients with a certain outcome (cases) for any previous exposures and comparing them with healthy individuals (controls). Examples of this type of study design include studies where we compare patients with a rare disease to patients without the disease to identify any previous exposures that those with the rare disease might have had in common. The advantages of case-control studies are that they offer a cost- and time-efficient means of recruiting a relatively large number of cases (e.g., from computer records), thus avoiding prolonged follow-up of large cohorts. The disadvantage, similar to that of a retrospective cohort study, is the potential absence of the historical data you need (Checkoway et al., 2007).

Cross-sectional study – This is the easiest type of study to conduct and the most commonly encountered in medical journals. They assess one point in time (and therefore are neither prospective nor retrospective). With these studies, you may compare disease prevalence among exposed and nonexposed groups, or among groups classified according to the level and type of exposure. The pros of a cross-sectional study are that it is feasible and easy to perform and that it allows you to find the prevalence of a medical condition (i.e., the proportion of a particular population found to be affected by it) (Pandis, 2014). However, it is often criticized for providing limited causal inference because exposure and health outcomes are usually assessed concurrently. In other words, such studies may be prone to a problem known as *reverse causation bias*—that is, the exposure status may be an effect of the disease rather than a cause.

Of note, it should be appreciated that some research questions may be investigated by more than one approach, although one design is usually the most preferable, either because it provides more direct causal evidence or due to its feasibility (based on the resources available). To clarify this point, say that we want to find out if hypertension predisposes a patient to stroke, and we want to use the same open data we accessed in the Chapter 1 tutorial; the Framingham Heart Study data from the Biologic Specimen and Data Repository Information Coordinating Center (BioLINCC) repository (www.biolincc.nhlbi.nih.gov). The research question is "Does hypertension predispose adults to stroke?" Let us apply different study designs to help answer this question. Moreover, you always should keep in mind the issue of confounding effect. The *confounding effect* is that the causality attributed to factor A is weakened if there is an alternative factor that causes both factor A and the outcome (Frank, 2000). An example is that when we are studying the effect of hypertension on stroke, the presence of diabetes in hypertensive patients is a confounding variable, and so it may have a contributing effect in the occurrence of stroke. Therefore, the effect of hypertension on stroke is confounded by the presence of diabetes in these patients.

If you do a prospective cohort study, you will recruit a large sample of the population with hypertension (the cohort) and follow up with them. You need to take into account the long period of time it might take to develop a stroke, which would make the study very costly. However, a potential benefit is that you may choose to screen the information (e.g., cardiac biomarkers) that you collect to identify confounding variables (e.g., genetic factors) that may lead to certain outcomes. So, this study design yields the best-quality data, but at a high cost.

Another alternative is to do a retrospective cohort study by reviewing the records of patients with hypertension (the cohort) and following their records to see if there are any stroke events. In this case, you will need to include a large number of hypertensive patients and find which ones have subsequently suffered a stroke. You also will not likely be able to include potential confounding variables (e.g., genetics) in this case because the data you are using already exists, and such data are not routinely obtained or available for hypertensive patients. So, although this type of study has a low cost, it is difficult to find the records needed for this number of patients.

You could also do a case-control study, where you would include patients with stroke (the cases) and review their records for the presence of risk factors. A case-control study needs a smaller sample than does a retrospective cohort study because you are directly including patients with stroke (the study group) and patients without stroke (the control group). However, as with retrospective cohort studies, you cannot include data on potential confounding variables that are not already routinely collected for hypertensive patients (e.g., genetics). So, although this type of study has a low cost and is easier to perform than a retrospective cohort study, its data will be of a lower quality than that of a prospective cohort study.

Finally, you could perform a cross-sectional study, where at one point in time, you would recruit stroke patients both with and without hypertension (i.e., the exposure), and study if there is any association between stroke and hypertension. Although it is also a relatively low-cost study, you cannot predict whether the exposure or the risk factor was the cause or the effect (i.e., whether stroke caused hypertension or vice versa).

To answer our research question, we can rank the study designs based on the quality of the resulting data and make a choice based on the feasibility of performing the study (e.g., availability of funding), starting with the best

Table 2.2   Comparing the Pros and Cons of the Four Main Observational Studies

Study Design	Quality of Collected Data	Feasibility (Cost and Burden)	Notes
Prospective cohort	++++	+	Well-suited for investigations of relatively short-term phenomena
Retrospective cohort	+++	++	Typically limited to mortality outcomes (due to availability of data)
Case-control	++	+++	More efficient and feasible relative to full cohort studies
Cross-sectional	+	++++	Most appropriate for studying relatively long-term conditions

Note: The number of plus signs is used here just to roughly estimate the difference in quality of collected data and feasibility between different study types.

quality-data designs: prospective cohort, retrospective cohort, and case-control (listed from best to worst). A cross-sectional study will also answer this particular research question, but you should pay attention to ensure that hypertension occurred before stroke, not vice versa. It is also always recommended that you search for previous studies with similar research questions and review their methodology to gain better insight into how to conduct yours, the best way to do the search would be on PubMed using the advanced search function with two keywords: the first for the exposure, the second for the outcome. Using PubMed to do your search provides you with the "best match" articles, as opposed to Google Scholar, which can retrieve less relevant results than you want (Anders and Evans, 2010).

Table 2.2 of this chapter provides a comparison of the levels of quality and feasibility among the four types of observational studies.

After choosing the best study design for your research question, it is now time to turn to your study details. You have to write every detail both in your study protocol initially, and the methodology part of the manuscript afterward, from how you chose your sample population to how you obtained information from them. You should write this description so clearly that if another scientist wanted to conduct the same research at another institution, that person could follow your steps and get the same results you did; this study characteristic is termed *reproducibility*, and will be explained next.

Reproducibility and bias

Note the following description from an editorial by Shafer and Dexter (2012):

> Greenberg and colleagues reveal an unexpected association between rain and the annual meeting of the Society of Pediatric Anesthesia (SPA). This is not a small effect! The odds ratio for rain on the first day of the SPA meeting is 2.63 when compared to historical controls. The finding is statistically significant, with a P value of 0.006. The authors conclude that "further investigation is warranted" and recommend that "SPA leadership may use [its] rainmaking ability to benefit drought-stricken areas."

These findings came about by pure chance, or the data might be made up altogether. There is very small chance of finding the same result for other researchers, so this is termed *irreproducible research*—a serious flaw. Reproducibility is related to the quality of the methodology, the better your reporting, the higher the reproducibility of your research will be. At this point, you should be aware of the term *bias* and how you can avoid it, thus producing highly reproducible research.

Bias is defined as any tendency that prevents the unprejudiced consideration of a question. In research, bias occurs when errors are introduced into sampling or testing by selecting or encouraging one outcome or answer over others (Pannucci and Wilkins, 2010). You should be aware that some degree of bias always exists in articles, even those published in high-impact journals; however, you should do your best to minimize the degree of bias in your study. To understand how bias can affect study outcomes, we cite observational studies prior to 1998 demonstrating that hormone replacement therapy decreases the risk of heart disease in postmenopausal women (Stampfer and Colditz, 1991). Recent, better-designed studies with minimal bias found the opposite effect for hormone replacement therapy on heart disease (Hulley et al., 2002). Bias may occur at any part of a study, from design to publication.

Table 2.3 summarizes the potential bias associated with each type of study design. The following biases are the most important in observational studies (Mann, 2003):

- *Design bias* is introduced when the researcher fails to take into account the inherent biases liable in most designs (refer to Table 2.3).
- *Selection/sampling bias* reflects the potential that the sample studied is not representative of the population that the study wants to focus on.
- *Procedural bias* occurs when an unfair amount of pressure is applied to the subjects, forcing them to complete their responses quickly.
- *Measurement bias* arises from errors in data collection and the process of measuring.
- *Interviewer bias* is introduced when the interviewer directly or indirectly guides the interviewee to certain answers. Interviewer bias is one of the most difficult research biases to avoid in many quantitative experiments when relying upon interviews.

Table 2.3   Comparing Bias Among the Four Main Observational Studies

Study Design	Potential Sources of Bias
Prospective cohort	Incomplete definition of exposure Lack of follow up (transfer bias) Data collection variability (if more than one data collector)
Retrospective cohort	Incomplete definition of exposure Missing data (recall bias) Data collection variability (if more than one data collector)
Case-control	Incomplete definition of cases Missing data (recall bias) Data collection variability (if more than one data collector)
Cross-sectional	Sampling techniques Data collection variability (if more than one data collector)

- *Response bias* is a type of bias where the subject consciously, or subconsciously, gives response that they think that the interviewer wants to hear.
- *Recall bias* occurs when people with the expected outcome of the study are more likely to remember certain antecedents, or exaggerate or minimize what they now consider to be risk factors.

The most important way to minimize bias in study design is to provide a clear definition of both exposures and outcomes, especially for both cohort and case-control studies. The use of objective definitions (preferably based on guidelines) with clear cutoff values for both exposures and outcomes can minimize interobserver variability. When collecting data for a retrospective study, double-check the records and validate information with other available references (e.g., contact the patient), especially when data collector encounters extreme values. Cross-sectional studies mainly involve collecting data such as answers to questionnaires, structured interviews, physical exams, laboratory or imaging data, and medical chart reviews. Because multiple individuals are gathering and entering data, standardized protocols for data collection, including training of study personnel, can minimize interobserver variability. Another good practice to reduce bias is to blind study personnel to the patient's exposure and outcome status, or, when this is not possible, the examiners measuring the outcomes should be different than those who had evaluated the exposures.

Life cycle of a study

After finishing your research and completing your manuscript, your aim is to publish it in a journal, a process that start with what is called submission, which is the request for publication. Choosing the right journal to submit to is essential for making sure that readers who are the most interested in your work will read it. There are three main aspects you should consider in any journal you want to submit to:

1. Its scope, which is usually found on the home page of the journal's website
2. Its strength, measured by several metrics, the most common of which is the impact factor, measured by the Science Citation Index (SCI)
3. Any special requirements, like article processing charges (APCs) upon acceptance (may reach up to $3,000 at the time of writing), or open-access policy

To choose the best journal for you, first look for journals with a scope that matches the aim of your study. You can either consult the journals where your referenced articles were published, or use search engines using your manuscript's keywords. Then, look at their impact factors and rank them in order (see Chapter 4 for more information on impact factors). Finally, make a note of journals with any special considerations (e.g., APCs). Now, you have created a list of potential journals with a scope matching your study's scope, sorted according to their strengths and their special requirements. You should also look through some articles in each potential journal in your list in order to get an idea about the strength of the articles they accept. For example, you should not send a questionnaire-based study done in one institution to the *New England Journal of Medicine (NEJM)*, a journal that mostly publishes randomized, controlled, multicenter trials. Another example is a study that we did on diagnosing cerebro-venous sinus thrombosis, where we searched for journals considering stroke and related diseases, with an average strength (impact factor between 1 and 3), and without APCs (as our research is not funded). In the end, we published it in the *Journal of Stroke and Cerebrovascular Diseases* (Al-Ryalat et al., 2016). The point is, aim to publish in the highest-impact journals that fit your scope and other requirements that are important to you.

When you decide to submit your manuscript to a specific journal, the manuscript begins its life cycle. Each journal has its own instructions for authors, which may typically be found on its website. Follow these instructions by adjusting your manuscript according to the journal's requirements (e.g., font size). After that, your manuscript will be ready to be sent, or submitted, to the journal. For more information about choosing a journal to submit to, see Chapter 4.

Figure 2.3 illustrates the manuscript life cycle, from when a manuscript is submitted for publication until it is released as an article. The manuscript will be picked up and assessed by an editor, after which the life cycle flowchart splits into two potential outcomes; the manuscript either goes back to the authors or moves on to the next stage. If the editor determines that the manuscript is unsuitable for publication, it will be sent back to the author(s) with a rejection letter. If the editor thinks the manuscript is suitable for the journal and is of sufficiently high quality, the manuscript will move on

Figure 2.3   Life cycle of a manuscript from submission to publishing.

to the next phase of its life cycle. This next phase is the peer review (also known as a *blinded peer review*), where the editor sends the manuscript to researchers from outside the journal. They read the manuscript carefully, ask questions, and recommend to the editor whether the article is ready to publish (Niederhauser et al., 2004). Peer reviewers are usually anonymous, although there have been some calls for more open peer reviews to increase transparency (Smith, 1999).

The editorial decision that you may receive is one of the following:

- Acceptance of the article without revisions (rare)
- Request for minor revisions (such as adjusting tables and figures, rewriting sections)
- Request for major revisions (which could involve repeating experiments)
- Rejection

Once the article has been assessed by the peer reviewers, it goes back to the authors to make edits (if necessary). If the article is rejected at this stage, you may submit it to a different journal, but keep in mind that you should never submit your manuscript to more than one journal at a time. If all adjustments are made and the reviewers confirm suitability of the article for publication in the journal, the editor sends a letter of acceptance to the authors. It is now in the production phase, the editor's job is complete, and the journal manager takes over. The role of the journal manager is to modify the paper to the journal's style so that it may be included in an upcoming issue. However, prior to the listing, the authors must approve what has been done with the article.

Essential ethics

Ethics in the context of research

When you think of the word *ethics,* the first thing that might come to your mind is the rules that define the difference between right and wrong, or between legal and illegal behavior. Indeed, the most common way of defining *ethics* is the norms for conduct that distinguish between acceptable and unacceptable behavior (Resnik, 2015). In the land of research, however, the term has a special additional meaning, mainly having to do with issues that might come up whenever human beings or animals are involved in an experiment.

Every researcher should be familiar with the major principles of ethics in research to ensure safe conduct of his or her study, as well as to avoid outright misconduct. Here, we will present the basic and common ethical principles that a researcher needs to know in order to professionally conduct and publish his or her first research, particularly with observational studies. Most of what we will present here are the product of what is known as the Declaration of Helsinki (World Medical Association, 2013). You might have heard of this before—it is frequently mentioned in the "Methods" section of research articles, in a statement such as "This research was conducted in accordance with the Declaration of Helsinki." When you finish this section, you will be able to add this ethical statement to your methodology, too.

Institutional review board (IRB)

Applying ethical principles of research begins with requesting the approval of your research from the institutional review board (IRB) of your institution. An IRB is an independent committee established to protect the rights and welfare of human participants in a study. Every institution that participates in studies must identify an IRB to review and approve them. The researchers make an IRB approval request, which includes a proposal of the study and all the documents that might be required by the board for the review. Remember that IRB is sometimes a required document to obtain access to the open data.

The major role of the committee is to review the full plan proposed for the study and to monitor and review any biomedical and behavioral research involving humans or animals, in order to ensure that it follows the ethical rules of research. The committee is also formally designated to accept (or reject) a research proposal after confirming that the research plans do not expose participants to unreasonable risks. The IRB also may terminate ongoing research if it is not being conducted in accordance with the stipulated requirements or an unexpected or serious risk to participants manifests itself. Depending on the institution, it may take an average of 1–3 months until the IRB informs you of its decision (Wolzt et al., 2009).

However, in some circumstances, you may ask for an expedited review for your research, which involves less waiting time for approval (Hirshon et al., 2002). An expedited review may be carried out by the IRB chairperson or by one or more experienced IRB members designated by the chairperson. Those reviewers may exercise all the authority of the IRB except that of rejecting the research. Rejection of proposed research in an expedited review still needs the full IRB. Circumstances that may qualify for an expedited review include, but are not limited to, studies of existing data, documents, records, pathological specimens, or diagnostic specimens (i.e., research studies done on open-access data).

During the IRB approval process, and over the course of the study, the researchers must not admit any participants to the study before receiving written IRB approval. Further, they must not make any changes to the study protocol without prior written approval from the board. The researcher must promptly report to the IRB any deviations from the protocol, changes that increase risk to participants, all adverse drug reactions, and any new information that may adversely affect the safety of participants.

Conflict of interest

A *conflict of interest* is any situation in which an individual author, group of authors, or corporation is in a position to exploit research results for a benefit, whether financial or some other type. Such competing interests may make it difficult for the authors and/or corporation to fulfill their duties objectively. Conflict of interest exists even if no unethical or improper act is committed. All authors should clearly disclose any conflicts of interest in their manuscript. If a reader suspects any undisclosed conflict of interest while reading a published article, he or she may contact the editor of the journal, and the editor has the right to investigate the concern and contact the authors or corporations directly to address it (ICMJE, 2004). Failure to disclose all conflicts of interest is considered scientific misconduct (Krimsky, 2007).

Authorship

The list of authors tells readers who did the work, and to what degree each author contributed to it, so that the right people get appropriate credit for their work. It is frequently recommended that all the contributors discuss and decide authorship in advance of starting their research. Researchers should discuss any publication credit (e.g., who should be the first author) before they even begin the project. Authorship is theoretically straightforward, but in the area of research, it can be a major cause of dispute among authors. The International Committee of Medical Journal Editors (ICMJE) has developed criteria for authorship that may be used by all journals, including those that distinguish authors from other contributors (ICMJE, 2004). ICMJE states that authorship is based on meeting *all* of the following criteria:

- Substantial contributions to the conception or design of the work; or the acquisition, analysis, or interpretation of data for the work
- Drafting the work or revising it critically for important intellectual content
- Final approval of the version to be published
- Agreement to be accountable for all aspects of the work in ensuring that questions related to the accuracy or integrity of any part of the work are appropriately investigated and resolved

Accordingly, regardless of one's scientific degree or career position, only those who played major roles in the research process should be listed as authors, even the most junior members of the team. On the other hand, contributions that are primarily technical do not warrant authorship.

Authorship not only involves the list of those who meet the criteria for being authors, but also the order in which the authors are listed and who is the corresponding author. Authors are generally listed based on their contribution to the study, the first author being the one who has made the greatest contribution. It is usually the most sought-after position because when your article is cited by another one, your article is usually referred to by the author named first (e.g., AlRyalat S et al.). Furthermore, a job promotion in many academic institutions may require the employee to be the first author in some of his or her publications. The last author, on the other hand, is usually the mentor or most senior member of the team. The corresponding author is the person who takes primary responsibility for communication with the journal during manuscript submission, the peer review, the publication process, and postpublication comments. The corresponding author may be any of the authors listed, and his or her contact information is published with the article so that readers may contact the research group with any feedback or other commentary. Although it is primarily an administrative role, some authors may equate it with seniority. Remember to choose contact details that are not likely to change in the near future.

As previously mentioned, others who did not contribute significantly to the work but have provided some support (e.g., statistical analysis) should be recognized in the "Acknowledgments" section. Currently, no rules exist to restrict the number of authors of an article; however, remember that having large numbers of authors usually increases the time it takes to prepare, review, and finalize a manuscript for publication.

Plagiarism

Plagiarism (derived from the Latin *plagiarius,* meaning "kidnapping") describes the act of stealing in research, although the concept does not exist in a legal sense. It is defined as the use of ideas, concepts, words, or structures without appropriately acknowledging the source in a setting where originality is expected. In simpler words, as a journal once replied to an author who committed plagiarism, "Your work is both good and original. Unfortunately, the parts which are good are not original, and the parts which are original are not good" (Pecorari, 2003). Nowadays, there are several online services that check for plagiarism, which work by searching the Internet for any similar exact phrases as the ones used in a study. To further understand what plagiarism is, we recommend that you test it yourself via the following steps, using a free plagiarism-checking service:

1. Copy a paragraph from a book or a Wikipedia page.
2. Go to https://www.quetext.com and paste the paragraph text into the text box on the page.
3. The report will show 100% similarities with the source you copied from.
4. Now, try to rephrase some words from the paragraph and recheck the plagiarism, where it will show that the text is clear from plagiarism.

In open data research, the issue of plagiarism is more difficult. As you are getting your data from another study, the method of its collection is usually the same for you and for the original study, which is why you need to pay attention to rewording the methodology in a way it won't be just a "copy-and-paste" from the original study. As we have stated, plagiarism itself may not have legal consequences (unless it involves other crimes such as copyright infringement or fraud). However, it is considered academic fraud and dishonesty, and many institutions and journals punish authors who commit plagiarism. Always cite your work properly by providing full references for what you write to avoid committing plagiarism. Similarly, self-plagiarism (i.e., rewriting your own previously published findings as new findings, without citing the original publication) is also considered plagiarism.

Informed consent

Obtaining informed consent might be required before starting to recruit participants in your study. This is usually obtained by a written document to the effect that a person voluntarily agrees to participate in a study after being fully informed about the study via verbal and/or written discussion with the staff. The informed consent form should contain all of the information that a person needs to know before making the decision about taking part in the study, including the following steps: invitation to participate, discussion of the research, assessment of the patient's understanding of the research, and agreement to participate in the study (Hardicre, 2014).

While documentation of informed consent is required in most clinical studies, the purpose of consent is to allow a person to waive his or her right to control certain aspects of or information about his or her life. But becoming part of a research study or disclosing some information in observational studies does not seriously undermine one's control or right to privacy (Gelinas et al., 2016). There are some occasions when a waiver or alteration of written informed consent is obtained from the IRB for study participants. The Code of Federal Regulations (CFR) has specified some situations where the requirements to obtain informed consent may be waived [see 45 CFR 46.116(c) and 45 CFR 46.116(d)]. These include the following: (1) the study involves no more than minimal risk to the subjects; (2) the waiver will not adversely affect the rights and welfare of the subjects; (3) the study cannot be carried out without the waiver; and (4) whenever appropriate, the subjects will be provided with any pertinent information after participation. These scenarios also apply whenever the recruitment of participants or participants' data comes from the Internet or any available open-access data resource. You should consult the IRB of your institution to find out when it is possible to waive the requirements of informed consent.

Confidentiality and privacy

Participant records and any other identifiable participant information must be kept confidential, especially in an era where medical records increasingly are stored electronically, and electronic information can be shared easily and widely (Kass et al., 2003). Even if the results of your study are published, participants' identities and their contact information must be protected. Remember that even if you are conducting research study using open-access data, you still must protect the confidentiality of your participants' information, and you are not allowed to contact any participant in the study personally. Some institutes, before giving you access to their data, require you to sign a form to ensure the privacy and confidentiality of their subjects and also may ask you to discard all their data documents immediately after finishing your study.

Retraction

Retraction is a mechanism for correcting the literature and alerting readers to publications that contain misinformation, so that readers know their results cannot be relied upon in their published form. The main purpose of retractions is to correct the literature, not to punish authors. Journal editors have the responsibility of ensuring the reliability of their scientific papers, implying that it will sometimes be necessary to retract an article. Unreliable data may result more commonly from honest, unintentional errors, but may also result from research misconduct and even fraud (Steen, 2011).

Retraction should be considered in cases of redundant publication (i.e., when authors publish the same results in several publications), plagiarism, unethical research, and failure to disclose a major conflict of interest that is likely to influence the recommendations of the study. Postpublication authorship disputes may sometimes result in the retraction of an article; however, if a change of authorship is required, but there is no reason to doubt the validity of the protocol and the findings of the study, retractions are usually not appropriate. For a more detailed guidance on retractions, refer to the Committee on Publication Ethics (COPE) retraction guidelines (publicationethics.org).

Protocol tutorial

From the Chapter 1 tutorial, we came up with an idea of to studying the difference between the body mass index (BMI) of patients with a previous history of cardiovascular events and the BMI of patients without such a history of cardiovascular events. The first step in the planning phase of a study (Figure 2.2) is to write your research question. The question should be operable, measurable, and fully descriptive, so our question based on this idea will be: "Is there a difference in BMI between patients with and patients without a previous history of cardiovascular events?"

The next step in the planning phase of the study will be to write a research protocol, which require a deep read. This step will focus on previous studies discussing the effect of stroke on BMI. Because our search will be more specific to this research question, we should use an advanced search in PubMed (as described in Chapter 1).

Here is a brief outline of how we will create the research protocol, the steps of which are summarized in this chapter. Then an example of a protocol developed from our research question will follow.

We are interested in the lifestyle habits of individuals who have suffered from cardiovascular disease. We are particularly interested in whether individuals who have suffered from cardiovascular disease later experience a change in Body Mass Index compared to their healthy counterparts with no previous history of cardiovascular disease.

1. Begin by writing your research question.
2. Do a thorough literature search using the advanced search function in PubMed, with the keywords "cardiovascular disease" AND "body mass index."
3. Write the background section of your protocol according to your search results.
4. Write the methodology.
5. Formulate your timetable according to your methodology.

Research protocol

Research question – Is there a difference in BMI between patients with and patients without a previous history of cardiovascular events?

Background – Cardiovascular disease generally leads to a wide range of adverse changes in the patient's life, from physical changes to loss of job security to anxiety (Von Känel et al., 2010). The American Heart Association (AHA) showed an improvement in the long-term survival of patients who suffered from acute myocardial infarction (MI) by reducing cardiovascular risk factors (Gomez-Marin et al., 1987), many of which were metabolic abnormalities. However, several previous reports have pointed out that patients who have suffered cardiovascular disease (particularly MI) are themselves at risk for developing metabolic abnormalities later. In a study that involved patients who had suffered an acute MI, 65% of patients were diagnosed with either impaired glucose tolerance or diabetes mellitus only 3 months after discharge (Norhammar et al., 2002).

It is well-known that high BMI is one of the risk factors for developing cardiovascular disease (Yusuf et al., 2005). In this study, we will discuss the opposite: the effect of previous history of cardiovascular disease on BMI.

Methodology – This study will be a secondary analysis of the Framingham Heart Study data, which we will obtain from BioLINCC (https://biolincc.nhlbi. nih.gov/studies/fhs/?q=heart). The Framingham Heart Study is a long-term, prospective cohort study on the etiology of cardiovascular disease among a population of free-living subjects in the community of Framingham, Massachusetts.

In our study, we will include patients with a previous history of cardiovascular disease and compare them to patients without a previous history of these conditions. Our primary goal is to compare the mean difference in BMI between patients with and those without a history of cardiovascular disease.

Timetable

1–2 weeks: Complete the application process and gain access to the Framingham Heart Study data.

2–3 weeks: Do a thorough literature review for relevant studies.

1–2 weeks: Complete statistical analyses.

2–3 weeks: Write the manuscript.

References

Alhawari HH, Al-Shelleh S, Alhawari HH, Al-Saudi A, Aljbour Al-Majali D, Al-Faris L, AlRyalat SA. Blood pressure and its association with gender, body mass index, smoking, and family history among university students. *International Journal of Hypertension*. 2018, 2018.

Al-Ryalat NT, AlRyalat SA, Malkawi LW, Al-Zeena EF, Al Najar MS, Hadidy AM. Factors affecting attenuation of dural sinuses on noncontrasted computed tomography scan. *Journal of Stroke and Cerebrovascular Diseases*. 2016 Oct 1;25(10):2559–2565.

AlRyalat SA, Malkawi L. International collaboration and openness in Jordanian research output: A 10-year publications feedback. *Publishing Research Quarterly*. 2018.

Anders ME, Evans DP. Comparison of PubMed and Google Scholar literature searches. *Respiratory Care*. 2010 May 1;55(5):578–583.

Checkoway H, Pearce N, Kriebel D. Selecting appropriate study designs to address specific research questions in occupational epidemiology. *Occupational and Environmental Medicine*. 2007 Sep 1;64(9):633–638.

Creswell JW. *Educational research. Planning, conducting, and evaluating quantitative and qualitative research.* 2008.

Dawber TR, Meadors GF, Moore Jr FE. Epidemiological approaches to heart disease: the Framingham Study. *American Journal of Public Health and the Nations Health*. 1951 Mar;41(3):279–286.

Dykacz JM. Prospective and retrospective studies. Wiley StatsRef: Statistics Reference Online, 2014 Apr 14.

Fisher JP, Jansen JA, Johnson PC, Mikos AG. *Guidelines for writing a research paper for publication*. Mary Ann Liebert, Inc. United states, New York, 2013.

Frank KA. Impact of a confounding variable on a regression coefficient. *Sociological Methods & Research*. 2000 Nov;29(2):147–194.

Gelinas L, Wertheimer A, Miller FG. When and why is research without consent permissible? *Hastings Center Report*. 2016 Mar 1;46(2):35–43.

Gomez-Marin O, Folsom AR, Kottke TE, Wu SC, Jacobs Jr DR, Gillum RF, Edlavitch SA, Blackburn H. Improvement in long-term survival among patients hospitalized with acute myocardial infarction, 1970 to 1980. *New England Journal of Medicine*. 1987 May 28;316(22):1353–1359.

Hardicre J. Valid informed consent in research: An introduction. *British Journal of Nursing*. 2014 Jun 12;23(11):564–567.

Hirshon JM, Krugman SD, Witting MD, Furuno JP, Limcangco MR, Perisse AR, Rasch EK. Variability in Institutional Review Board Assessment of Minimal-risk Research. *Academic Emergency Medicine*. 2002 Dec 1;9(12):1417–1420.

Huang WC, Levey AS, Serio AM, Snyder M, Vickers AJ, Raj GV, Scardino PT, Russo P. Chronic kidney disease after nephrectomy in patients with renal cortical tumours: A retrospective cohort study. *The Lancet Oncology*. 2006 Sep 1;7(9):735–740.

Hulley S, Furberg C, Barrett-Connor E, Cauley J, Grady D, Haskell W, Knopp R et al. Noncardiovascular disease outcomes during 6.8 years of hormone therapy: Heart and Estrogen/progestin Replacement Study follow-up (HERS II). *Jama*. 2002 Jul 3;288(1):58–64.

International Committee of Medical Journal Editors (ICMJE). Uniform requirements for manuscripts submitted to biomedical journals: writing and editing for biomedical publication. *Haematologica*. 2004 Mar 1;89(3):264.

Kass NE, Natowicz MR, Hull SC, Faden RR, Plantinga L, Gostin LO, Slutsman J. The use of medical records in research: What do patients want? *The Journal of Law, Medicine & Ethics*. 2003 Sep;31(3):429–433.

Krimsky S. When conflict-of-interest is a factor in scientific misconduct. *Med. & L.* 2007;26:447.

Leedy PD, Ormrod J. *Practical Research–Planning and Design*. 8th ed. Upper Saddle River, NJ: Pearson; 2005.

Mann CJ. Observational research methods. Research design II: Cohort, cross sectional, and case-control studies. *Emergency Medicine Journal*. 2003 Jan 1;20(1):54–60.

McNair P, Lewis G. Levels of evidence in medicine. *International Journal of Sports Physical Therapy*. 2012 Oct;7(5):474.

Melese T, Lin SM, Chang JL, Cohen NH. Open innovation networks between academia and industry: An imperative for breakthrough therapies. *Nature Medicine*. 2009 May 1;15(5):502.

Moeen F, Mumtaz R, Khan A. Writing a Research Proposal. *Pakistan Oral & Dental Journal.* 2008 June;28(1):145–152.

Moses H, Dorsey ER, Matheson DH, Thier SO. Financial anatomy of biomedical research. *JAMA.* 2005 Sep 21;294(11):1333–1342.

Niederhauser DS, Wetzel K, Lindstrom DL. From manuscript to article: Publishing educational technology research. *Contemporary Issues in Technology and Teacher Education.* 2004;4(2):89–136.

Norhammar A, Tenerz Å, Nilsson G, Hamsten A, Efendíc S, Rydén L, Malmberg K. Glucose metabolism in patients with acute myocardial infarction and no previous diagnosis of diabetes mellitus: A prospective study. *Lancet.* 2002 Jun 22;359(9324):2140–2144.

Omair A. Selecting the appropriate study design for your research: Descriptive study designs. *Journal of Health Specialties.* 2015 Jul 1;3(3):153.

Pandis N. Cross-sectional studies. *American Journal of Orthodontics and Dentofacial Orthopedics.* 2014 Jul 1;146(1):127–129.

Pannucci CJ, Wilkins EG. Identifying and avoiding bias in research. *Plastic and Reconstructive Surgery.* 2010 Aug;126(2):619.

Pecorari D. Good and original: Plagiarism and patchwriting in academic second-language writing. *Journal of Second Language Writing.* 2003 Dec 1;12(4):317–345.

Rawat S, Meena S. Publish or perish: where are we heading? *Journal of Research in Medical Sciences: The Official Journal of Isfahan University of Medical Sciences.* 2014 Feb;19(2):87.

Resnik DB. What is ethics in research and why is it important? *Inideas.* 2015 Dec 1.

Ressing M, Blettner M, Klug SJ. Systematic literature reviews and meta-analyses: Part 6 of a series on evaluation of scientific publications. *Deutsches Ärzteblatt International.* 2009 Jul;106(27):456.

Röhrig B, Du Prel JB, Blettner M. Study design in medical research: Part 2 of a series on the evaluation of scientific publications. *Deutsches Ärzteblatt International.* 2009 Mar;106(11):184.

Shafer SL, Dexter F. Publication bias, retrospective bias, and reproducibility of significant results in observational studies. *Anesthesia & Analgesia.* 2012 May 1;114(5):931–932.

Smith R. Opening up BMJ peer review: A beginning that should lead to complete transparency. *BMJ: British Medical Journal.* 1999 Jan 2; 318(7175):4.

Stampfer MJ, Colditz GA. Estrogen replacement therapy and coronary heart disease: A quantitative assessment of the epidemiologic evidence. *Preventive Medicine.* 1991 Jan 1;20(1):47–63.

Steen RG. Retractions in the scientific literature: Is the incidence of research fraud increasing? *Journal of Medical Ethics.* 2011 Apr 1; 37(4):249–253.

Troyka LQ. *Simon and Shuster handbook for writers.* 3rd ed. Upper Saddle River, NJ: Prentice-Hall, 1993.

Vernon MM, Balas EA, Momani S. Are university rankings useful to improve research? A systematic review. *PloS One.* 2018 Mar 7;13(3):e0193762.

Vetter TR, Mascha EJ. Defining the Primary Outcomes and Justifying Secondary Outcomes of a Study: Usually, the Fewer, the Better. *Anesthesia & Analgesia.* 2017 Aug 1;125(2):678–681.

Von Känel R, Kraemer B, Saner H, Schmid JP, Abbas CC, Begré S. Posttraumatic stress disorder and dyslipidemia: Previous research and novel findings from patients with PTSD caused by myocardial infarction. *World Journal of Biological Psychiatry.* 2010 Jan 1;11(2):141–147.

Wolzt M, Druml C, Leitner D, Singer EA. Protocols in expedited review: Tackling the workload of ethics committees. *Intensive Care Medicine.* 2009 Apr 1;35(4):613–615.

World Medical Association. Declaration of HelsinkiEthical Principles for Medical Research Involving Human Subjects. *JAMA.* 2013;310(20):2191–4, doi: 10.1001/jama.2013.281053

Yusuf S, Hawken S, Ounpuu S, Bautista L, Franzosi MG, Commerford P, Lang CC et al. Obesity and the risk of myocardial infarction in 27,000 participants from 52 countries: A case-control study. *Lancet.* 2005 Nov 5;366(9497):1640–1649.

3
Statistics You Need

Saif Aldeen Saleh AlRyalat

Essential statistical knowledge

This section is intended to provide you with a background of some points that might have slipped your mind as you read or perform research. First, we will provide an insight of what a p-value is and how to put it in context. Then we will go to the statistical concept of hypothesis testing and look at it from different angles. Next, the issue of calculating sample size and how to judge whether a sample is representative will be discussed. Finally, we will examine different types of study design from the statistical perspective.

p-value

The term *p-value* is short for "probability value"; it measures how well the sample data support your argument. Therefore, you must already have a hypothesis in mind if you want to use a p-value. The American Statistical Association (ASA) defined p-value as "the probability under a specified statistical model that a statistical summary of the data (e.g., the sample mean deference between two compared groups) would be equal to or more extreme than its observed value" (Wasserstein and Lazar, 2016). One of the main issues that arises about the p-value is overinterpretation, so a scientist could look at the p-value of 0.04 and say that "there was only 4% chance of the results being false." However, a p-value does not measure the probability that a particular hypothesis is true, or the probability that data were produced by random chance alone; rather, it only indicates how your data are compatible with the hypothesis.

The idea here is that you should have your hypothesis well developed (as described next), and then use the right test on the right sample to see if your data support your hypothesis by getting a p-value below the threshold of significance (i.e., 0.05). Using tests haphazardly until you get a p-value less than 0.05 isn't the right way to do it. Rather, you always need to use the test in its context to get a p-value, and then interpret that in the right way, so do the test in the following circumstances:

1. You have a valid hypothesis.
2. You have an adequate representative sample.
3. You choose the right test for your sample and hypothesis.

Only then can you conclude confidently whether your finding is significant or not. To further clarify the point, the following example was published in an article that was one of the most highly viewed in nature (Nuzzo, 2014):

> For a brief moment in 2010, Matt Motyl was on the brink of scientific glory: he had discovered that extremists quite literally see the world in black and white. The results were "plain as day," recalls Motyl, a psychology PhD student at the University of Virginia in Charlottesville. Data from a study of nearly 2,000 people seemed to show that political moderates saw shades of grey more accurately than did either left-wing or right-wing extremists. The p-value, a common index for the strength of evidence, was 0.01—usually interpreted as "very significant." Publication in a high-impact journal seemed within Motyl's grasp.
>
> But then reality intervened. Sensitive to controversies over reproducibility, Motyl and his adviser, Brian Nosek, decided to replicate the study. With extra data, the p-value came out as 0.59—not even close to the conventional level of significance, 0.05. The effect had disappeared, and with it, Motyl's dreams of youthful fame.

This brings us to the issue of replicability and reproducibility, as significant results mean that you can get the same result when you follow the same methodology. In the previous example, the author had found a significant p-value without being able to get the same results when he repeated the same methods (nonreproducible results), and that illustrates why we need to follow these three conditions to get a p-value that can assure us that we have a reproducible result.

Finally, it should be noted that a p-value never measures the size of an effect or the importance of a result. When a p-value for a tested hypothesis is 0.05 and for another hypothesis is 0.0001, this does not mean that the first hypothesis is stronger or more important than the other. Moreover, you should never report a p-value without reporting the setting in which it was used, as proper influence requires full reporting and transparency, as we will discuss in Chapter 4.

It is almost impossible to study the whole country's population every time we want to do a study on this population (except for the U.S. census, performed every 10 years). The alternative is to study a sample from the population that one can confidently use to represent the whole population. Such is the role of statistical analysis, which determines the sample that will represent your population and the generalizability of your findings to the whole population.

Determining your sample population

When you want to answer your research question (i.e., test your hypothesis), you need a sample that is representative of your target population. When you want to apply for a grant to study certain characteristics in a population, the first thing the evaluators will look at is the sample you need, to make sure that you would not be wasting their money. But how do you choose a representative sample? What sample size do you need?

Representative sampling – *Sampling* is the process of selecting representative elements of a population for inclusion in a study that allows inferences and generalizations about the population without actually examining each element in the population. The first step is to define the target population. For example, if you are studying a condition in a city, then your target population is the city's population; but if you are studying a condition in an entire country, then your target population would be the country's population. After defining your target population, there are two main sampling techniques to follow: probability and nonprobability sampling (Kothari, 2004). *Probability sampling* (e.g., simple random sampling) is where elements are chosen randomly and every element has an equal and independent chance of being selected. Although this type of sampling is laborious, it is a better representation of your target population. On the other hand, *nonprobability sampling* (e.g., convenience sampling) is where elements are chosen by nonrandom methods and there is no way of ensuring that every element has an equal chance of being selected. Although this type of sampling is less rigorous, it is also less representative of your target population.

Sample size – In simple terms, the goal of calculating sample size is to include the smallest sample needed to get reliable results. The idea behind a proper sample size calculation is that, for example, we do not want to include 1,000 patients in a study if 100 is enough to detect the effect we are looking for, or vice versa. To determine the sample size, you should be aware of the following concepts (Button et al., 2013): *Statistical power* is the probability that a test will define the presence of an effect when the effect is present. Correctly finding an effect depends on two main factors: the magnitude of this effect (effect size), and having enough sample to detect this effect (sample size). Effect size quantifies the size of the difference or association between two groups. For example, the mean difference in length between men and women from the same ethnicity is relatively large, so it is easy to detect the effect size in length between men and women. To sum up, you need to maintain a high power level for your study. To do so, you need to increase the sample size when the effect size is small in order to detect it, and you can use a smaller sample size when the effect size is large. To perform a power analysis, we recommend the use of the freely available G*Power tool (http://www.gpower.hhu.de/en.html). You also can read more about power analysis in the available references given by Faul et al. (2009).

Hypothesis testing

In any statistical analysis, there are three main components that you need to test your hypothesis and answer your research question: the hypothesis we are testing, the variables that we are testing, and the tests themselves.

The hypothesis – The first step in statistical testing is formulating your hypothesis. A *hypothesis* is the statistical version of your initial research question. Let us take the question formulated in the previous chapter: "Is there a difference in BMI between patients with and patients without a previous history of cardiovascular events?" From this question, we should derive our statistical hypothesis in two forms:

- *The null hypothesis (H$_0$):* The answer to your question is negative (i.e., "There is *no* difference in mean BMI between patients with and patients without previous history of cardiovascular events"). The null hypothesis says that the findings are the result of chance or random factors.
- *The alternative hypothesis (H$_1$):* The answer to your question is positive (i.e., "There *is* a difference in mean BMI between patients with and patients without previous history of cardiovascular events"). The alternative hypothesis is what you hope to prove.

This type of hypothesis detailing is only used for performing statistical analyses. A hypothesis test uses collected sample data to determine whether you can reject the null hypothesis and accept the alternative (i.e., find that the answer to your research question is positive) or vice versa (i.e., find that the answer to your research question is negative).

Regarding the null hypothesis, two basic forms of error are recognized and should be avoided (Gonzalez-Mayo et al., 2015):

- *Type I error:* This occurs when the null hypothesis (H$_0$) is actually true, but is erroneously rejected. In other words, in our example, when the answer to the research question was actually negative, but your data found a positive answer. A type I error may be described as a "false positive." It is denoted as the alpha level, or significance level, and is usually set to 0.05 (5%). It means that we accept a 5% probability of incorrectly rejecting the null hypothesis.
- *Type II error:* This occurs when the null hypothesis (H$_0$) is actually false, but erroneously fails to be rejected. In other words, in our example, when the answer to the research question was actually "yes," but your data found "no" as an answer. A type II error is mainly due to an insufficient sample size to represent your target population. A type II error may be described as a "false negative." It is denoted as the beta level, and is related to the power of a test (see the earlier discussion of sample size). If you fail to reject the null hypothesis due to insufficient sample size to detect the effect size (low power), you have a type II error.

The variables – There are two main ways to classify variables (Peat & Barton, 2008). The first is according to its nature (categorical or continuous). The second is according to its role (independent or dependent).

Categorical variables are also known as discrete or qualitative variables, and they describe the quality. Categorical variables may be further categorized as either nominal or ordinal. Nominal variables have two (*dichotomous* or *binary*) or more (*multinominal*) categories. They do not have an intrinsic order (e.g., countries like Jordan, the United States, and the United Kingdom, which can be listed in any order). Ordinal variables, on the other hand, have categories that may be ordered or ranked (e.g., the Likert scale: with Very good, Good, Average, Bad, and Very bad rankings).

In contrast, continuous variables (also known as quantitative variables) describe data that can be scaled. Continuous variables may be further categorized as either interval or ratio variables according to the meaning of the zero. It is worth knowing about all these types of variables, although many software programs usually use only three types: nominal, ordinal, and scale (i.e., continuous), without differentiating between interval or ratio variables.

We may also categorize a variable according to its role. An independent variable (also known as an *experimental* or *predictor variable*) is a variable that is being manipulated in an experiment in order to observe the effect on a dependent variable (also known as an *outcome variable*).

The tests – The two main types of tests for statistical analysis are descriptive statistics and inferential statistics. Descriptive statistics are solely concerned with the properties of the observed data. Inferential statistics, on the other hand, make assumptions about the properties of the population.

Descriptive statistics are used to describe the basic features of the data in a study. They provide simple summaries about a sample to help us to simplify large amounts of data in a sensible way. There are three major characteristics of a single variable that we tend to look at in descriptive statistics: distribution, central tendency, and dispersion. *Distribution* is a summary of the frequency of individual values or ranges of values for a variable. *Central tendency* is an estimate of the center of distribution of values by using mean, median, and mode. *Dispersion* is the spread of the values around the central tendency, represented by range, variance, and standard deviation (SD). These will be further discussed later in this chapter in the section "Statistical Analyses."

Inferential statistics, on the other hand, are used to make conclusions about the population that our sample represents. In other words, it would not be necessary to use inferential statistics if you had included the whole target population. But because we can rarely observe and study entire populations, we try to select a sample that is representative of the entire population. We may then generalize the results from that sample to the target population. We depend on inferential statistics to test our hypothesis, which we will detail in the next section, with a tutorial given at the end of the chapter to help further understanding.

Statistics and study design

We have discussed different study designs in Chapter 2. Briefly, there are three main types of study design: cohort studies (either retrospective or prospective), case-control studies, and cross-sectional studies. But what type of information can we expect from each study? This question will be discussed next.

Cohort study – The cohort type of study follows a group of patients with certain risk factors to observe disease. It assesses a single risk factor that may be related to many diseases. The outcomes expected from this type of study are relative risk (RR), attributable risk (AR), and incidence (as discussed later). *Relative risk (RR)* is expressed by the comparative probability question, "How much more likely is the exposed person going to get the disease compared to the nonexposed?" *Attributable risk (AR)* addresses the comparative probability question, "How many more cases in an exposed group compared to nonexposed group?" (Bigby, 2000).

Case-control study – It is concerned with reviewing risk factors in patients who already have the disease. It assesses many potential risk factors for a single disease. The outcome is the odds ratio (OR). The OR looks at the increased odds of getting a disease with exposure to a risk factor as opposed to no exposure to that factor (Le & Eberly, 2016).

Cross-sectional study – It assesses the association between the risk factor and the disease at a single time point. The outcomes are disease prevalence (as discussed later in this chapter) and associations (Le & Eberly, 2016).

Now, let's detail the difference between incidence and prevalence. *Incidence* is a rate describing the proportion of new cases in a defined population in a specific period of time. We usually use incidence to describe acute diseases. *Prevalence* is a rate describing the proportion of cases in a defined population in a specific period of time. We usually use prevalence to describe chronic diseases.

In a cohort study, a group of individuals exposed to an agent are compared with a control group of individuals who were not exposed. Both groups are observed for a prespecified time period until the event of interest occurs. The association of exposure and outcome is expressed as RR, calculated by dividing the incidence rate of the exposed group by the incidence rate of the unexposed group.

In a case-control study, the association between exposure and outcome is followed by predefining cases and their matching controls. Therefore, we begin by marking the cases (the group with the outcome we want to study) and the control (the group without the outcome of interest) and review their records retrospectively. We use this study to find the OR, which is calculated by dividing the odds of exposure for the cases by the odds of exposure for the controls.

Using cross-sectional studies is easier because you are only finding associations between your variables according to your hypothesis.

Statistical analyses

More than 100 statistical tests may be used to analyze data. You do not need to know all of these statistical tests; you just need to know the tests you will use in your study and master them. Some of them are commonly used, and you probably have already heard of them.

When we are talking about open data, you already have your ready-to-use sample by default, so you do not have to perform a sample size calculation or design or validate a questionnaire for data collection. Even so, you need to know how to analyze your raw data to investigate your hypothesis and get the results you need. You always need to begin with descriptive statistics to give the reader an idea of your study. Most researchers are familiar with descriptive statistics (e.g., mean and median). However, if you are going to analyze the data yourself, you also need to know about inferential statistics to test your hypothesis. The main focus of this section is to discuss inferential statistics. However, you should always consult an experienced statistician about your analysis, especially if this is your first time.

Descriptive statistics

Descriptive statistics are used to give details about the basic features of data in a study. They provide simple summaries about the sample. The more information you provide in your description, the stronger your study will be. Generally, there are three main characteristics you should describe: distribution, central tendency, and dispersion (Marshall & Boggis, 2016).

Distribution is simply the frequency (or percentage) of values for a variable (e.g., the number of men and women). The simplest distribution would list every value of a variable and the number of persons who had each value. It is preferable to describe your data distribution in frequency tables in order to better communicate your results.

Central tendency is an estimate of the center of distribution. Measures of central tendency are the mean (or average, which is the sum of your values divided by their count), the median (the score in the exact middle of your values, where half of the scores above it and the other half below it), and the mode (the most frequently occurring value).

Dispersion is the spread of values around the central tendency. The two common measures of dispersion are standard deviation (SD) and range. *SD* shows the relationship that a set of scores has to the mean of the sample, and *range* is the difference between the highest and lowest values. Always try to report the mean and SD together in the form of "mean ± SD," as the mean alone does not provide the full picture about your data.

Inferential statistics

For any particular set of data, there might be dozens of tests that can be used to analyze it. But as an early career researcher, you will need few tests to get your results (Nayak & Hazra, 2011). This section will follow the layout of the diagram shown in Figure 3.1; it will guide you on choosing which test you will need, the conditions under which you will use it (assumptions), and how to report on your data eventually. We use SPSS in this chapter (IBM Corp., 2012), but you may use other software, as the concept is similar. SPSS itself has different software versions, so you will need to refer to the software user manual to execute each test. We believe that statistics is not a step-by-step calculation to reach a conclusion; rather, statistics should be influenced by the researcher's judgment to reach not only a statistically sound result, but also a scientifically sound conclusion. Moreover, as we stated at the beginning of this chapter, you can easily manipulate statistics to reach your conclusion, but you shouldn't. This fact has been appreciated

Group Differences

Associations

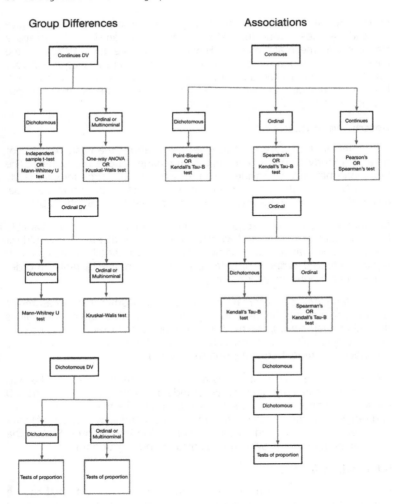

Figure 3.1   These flowcharts will guide you on choosing which test you will need. DV: dependent variable, IV: independent variable.

for a long time—indeed, in 1949, Evan Esar said that "statistics is the only science that enables different experts using the same figures to draw different conclusions" (Esar, 1953).

Figure 3.1 shows the tests that are the most commonly used and that you will most likely need. The diagram depicts two main pathways: group differences and associations. *Group differences* analyze differences between two or more groups, such as mean body mass index (BMI) between men and women, whereas *associations* analyze potential relationships between two or more variables (e.g., when BMI increases, so does the risk of cardiovascular disease). Starting from the top of the diagram, work your way down to the bottom. In each pathway, you will narrow down the tests based on the nature of your dependent variable (dichotomous, ordinal, or continuous, as discussed in subsection 3.1.3) Then, you will choose the nature of your independent variable (dichotomous, multinomial, ordinal, or continuous). Finally, you will choose from several statistical tests based on the specific assumptions for each test.

To simplify the concept of choosing the most suitable test for your hypothesis, imagine that the statistical tests are processing machines for raw material (data) to produce the final product (results or outcomes). The most important part here is to choose the best machine, which means taking into account three main considerations (Figure 3.2):

1. The machine must fit your factory type (group design).
2. The machine must be compatible with the raw material you have (types of variables).

Figure 3.2   The process of choosing the right statistical test for your raw data to get the right result can be imagined as choosing the right machine that can process your raw material to get the desired end product, where you can't choose any machine haphazardly.

3. The machine must have its own conditions for the raw material that need to be met in order to work properly (assumptions).

Group design – Understanding your study design is the first step to choosing the most suitable test for your hypothesis. With using open data, this understanding comes from reading about the design of the original study your data came from and the variables it had. This understanding will enable you to recognize the relationships among your variables. From a statistical point of view, you need to differentiate your study groups as either *between subjects* or *within subjects*. Once you do so, you may choose which column you will proceed with in the diagram (i.e., group differences or associations).

Between subjects means that there must be different participants in each group, with no participant being in more than one group (e.g., the difference in weight between men and women). This is also known as *independence of observations*. For this design, you may do the group differences tests (Figure 3.1).

Within subjects (also known as a *repeated measure design*) means that all cases are present in all the groups. In other words, the participants are either the same individuals tested at two time points, or under two conditions (e.g., for the same participants, their difference in weight now and one year later). Alternatively, you could have two groups of participants that have been matched (or paired) on one or more characteristics (e.g., gender). This simply means that the value of a participant should be compared to the value of its matched pair. For this design, you may do the association tests (Figure 3.1).

Types of variables – Now you know which group of tests to use, but which of those tests should you run? The answer is based on the nature of your variables. These variables are mostly dependent variables, although in associations, dependent variables do not have to be differentiated strictly from independent variables, which is why we didn't specify the nature of the variable in the association column. After choosing your type of dependent variable, proceed down the diagram according to your independent variable. Here, you will end up with two or more tests. Choosing the best test from these will be based on the assumptions that you make, which is discussed next.

Assumptions – Each statistical test needs certain assumptions to be met to perform efficiently. These assumptions are either related to your study design and variables or related to the nature of the data. We will not mention the assumptions related to study design and variables, as you should already have encountered them if you followed the steps shown in Figure 3.1. Assumptions related to the nature of the data, on the other hand, will be detailed shortly. Before choosing a statistical test, your data first should be tested to see which assumptions they meet. Note that on the diagram, the numbers next to each test indicate which assumptions should be met. The following assumptions can be tested using SPSS software, and the steps will depend on the version of the program you are using, so refer to the SPSS manual when using it. A tutorial at the end of this chapter will simulate how to do most of these assumptions.

Before we discuss the assumptions, you should be aware that there are two types of statistical tests: parametric tests, which are tests that should be done on continuous variables and require assumptions of certain distributions of data (usually normal distributions) and homogeneity of variance; and nonparametric tests do not assume normal distribution of data (Garson, 2012).

Assumptions related to the nature of data (Garson, 2012; Cummiskey et al., 2012; Marshall & Boggis, 2016) are discussed next.

1. *No outliers: Outliers* are simply extreme values; they can be defined statistically as values that are away from the mean by either 1.5 or 3 box-lengths. Tested visually via box plots, the outliers are the points (observations) that are more than 1.5 box-lengths from the edge of a box plot. If the point is more than 3 box-lengths from the edge, it is termed an *extreme outlier.* If you have outliers, you will need to manage them.

   Management: First, check to see if you made a mistake during data measurement or entry and correct the observations. Next, either exclude the extreme outliers and run the test with the remaining outliers, or go for the nonparametric test option, as will be explained in our discussion of each statistical test later in this chapter (see also Figure 3.1). Alternatively, you may do data transformation, but this is not recommended for beginners (Conover & Iman, 1981). Whatever your choice, you need to report it in the "Methodology" section of your paper.

2. *Normal distribution:* Normal distribution assumes that all observations are symmetrically distributed in a bell-shaped form around the mean (e.g., observed on a histogram, a plot that shows the frequency of each variable). This is tested by observing a bell-shaped distribution on histograms, or using the Shapiro-Wilk test for normality. The test yields a significant level by testing the null hypothesis, stating that variables are normally distributed. If $p > 0.05$, you accept the null hypothesis (i.e., the normal distribution assumption is met). If $p < 0.05$, then you reject the null hypothesis (i.e., the normal distribution assumption is not met), and you will need to manage further.

   Management: Go for the nonparametric version of the test (see Figure 3.1), as will be explained later, or do data transformation (not recommended for beginners).

3. *Similar distribution:* Similar distribution is tested visually via histograms. If all of your variables' distributions are similar in shape (regardless of their exact locations on the graph), you should go for median reporting (as explained in the tests covered later in the chapter). If they are not similar in shape, then you need further management.

   Management: Report using mean ranks, as explained later in this chapter (you might need to go for the other, nonparametric test according to Figure 3.1).

4. *Homogeneity of variance:* Variance is simply how far the values are dispersed from the mean. The homogeneity of variance assumption requires the variance in each populations to be equal. This is tested via Levene's test of equality of variance, which shows how far the values are spread from their average. The test yields a significant level by testing the null hypothesis, stating that the variables have equal variances. If $p > 0.05$, report the results for equal variances. If $p > 0.05$, report the results for unequal variances.

5. *Sample size:* Sample size is tested via frequency tables, which are tables showing the frequency of each variable. You need to make sure that the expected count for each variable tested in each cell of the frequency table is ≥ 5. If it is not, you need to manage further.

   Management: Collapse the variables to increase the count by combining variables with similar characteristics into a single variable (e.g., combining variables describing medical subspecialties like neurology and cardiology into one variable named "Medical subspecialties," having the count of all the variables), or go for the other test, as illustrated in Figure 3.1 and described later in this chapter.

6. *Linear relationship:* The linear relationship is tested visually via a *scatterplot*, which is a graph showing the distribution of all the observations, by denoting each observation as a dot, by plotting each two variables on the *x*- and *y*-axes. You should observe a linear relationship. If not, you need to manage further.

   Management: Go for the nonparametric version of the test (Figure 3.1), or do data transformation (not recommended for beginners).

7. *Monotonic relationship:* This is tested visually via a scatterplot, by plotting the variables on the *x*- and *y*-axes as described previously for the linear relationship assumption. You should observe a monotonic relationship (i.e., if one variable increases, the other variable increases or decreases, regardless of the linearity of the relationship). If the relationship is not monotonic, you need to manage further.

   Management: Go for the other, nonparametric test (according to Figure 3.1).

Group differences – Group difference tests are used to find if there are differences between groups (Laerd Statistics, 2017).

Independent samples *t* test – You should use the independent samples *t* test when you have a continuous dependent variable and a dichotomous independent variable and you want to compare the difference in means of the dependent variables between the two independent groups. Your null hypothesis (accepted if $p > 0.05$) is that the means of your two groups are equal, and your alternative hypothesis is that the means are different (accepted if $p < 0.05$).

Assumptions – The three main assumptions for this test are assumption 1 (no outliers), assumption 2 (normal distribution, for the continuous dependent variable for both dichotomous groups), and assumption 4 (homogeneity of variance).

The independent samples t test is a robust test, so that some violation of the normality assumption can be tolerated so that the test will still provide valid results. Therefore, you can accept an approximately normal distribution.

The results of Levene's test (to test assumption 4) are presented in the first section of the independent samples test table in SPSS, as demonstrated in the tutorial at the end of this chapter. If the population variance of both groups is equal, Levene's test will return $p > 0.05$, indicating that you have met the assumption of homogeneity of variances. You would report the results in SPSS as "equal variances assumed." However, if the test returns $p < 0.05$, the population variances are unequal, and you have violated the assumption of homogeneity of variances. SPSS already provides you with the result if the homogeneity of variance assumption is violated (in case Levene's test reveals a significant level of <0.05), these results are the product of a test called *Welch's test*.

Reporting – Begin by reporting that you met the outlier and normality assumptions, and state the significant-level Levene's test (for assumption 4). On SPSS, one of the output tables of the test will present the descriptive statistics, of which you should report the mean and SD (mean $\pm$ SD; mean age BMI of 32 ± 3) for your groups. Then, state the difference in means between the lower and upper bounds (difference in mean, lower bound to upper bound, are present in the test result's table) representing the 95% confidence interval, followed by the p-value. If Levene's test reveals a significant level of <0.05, you would then report the results in SPSS for "equal variances not assumed" in a similar manner, and you just need to state that the homogeneity of variance assumption was violated, so you will report the results of Welch's test.

Mann-Whitney U test – The Mann-Whitney U test can be thought of as the nonparametric version of the independent samples t test. You should use this test to find if there is a significant difference when you have a continuous or ordinal dependent variable or a dichotomous independent variable. You would consider using this test instead of the independent samples t test if your data failed the normal distribution assumption (assumption 2).

Your null hypothesis ($p > 0.05$) is that both distributions are equal. Your alternative hypothesis will depend on whether your data meet assumption 3 (similar distribution). If they do, then your alternative hypothesis ($p < 0.05$) is that the medians of the two groups are different. If you fail assumption 3, your alternative hypothesis ($p < 0.05$) is that the mean ranks of the two groups are different.

Assumptions – The assumption of this test is assumption 3 (similar distribution). If you meet this assumption, you may report your results more easily by using the medians. But if you violate this assumption, then you will need to report your results using the more uncommon concept of mean ranks.

Reporting
- *If reporting medians:* Begin by stating that you found similar distributions using a box plot. Then create a descriptive table that includes medians for your groups. Then state the difference in medians followed by the *p*-value.
- *If reporting mean ranks:* Begin by stating that you found different distributions using a box plot. Report that the mean rank of one group is more/less than that of the other group, followed by the *p*-value.

One-way analysis of variance (ANOVA) – You may consider one-way analysis of variance (ANOVA) when you want to find out if there is a significant difference between a continuous dependent variable and an ordinal (or multinominal) independent variable. Moreover, sometimes you might be interested in finding the exact difference between the items of the ordinal variable (e.g., you found that there is a significant difference between mean BMI and ethnicity using one-way ANOVA, but you are interested in finding in which ethnicity the difference is most significant—between whites and African Americans or white and Middle Eastern ethnicities?). In this case, you can use a post hoc test (a Latin word meaning "after truth") when you find that there is a difference between groups, but you want to know exactly where the difference is.

Your null hypothesis ($p > 0.05$) is that the means of the groups of the ordinal (or multinominal) variable are equal. Your alternative hypothesis ($p < 0.05$) is that the means are different.

Assumptions – The three main assumptions for this test are assumptions 1 (no outliers), 2 (normal distribution, for the continuous variable for all ordinal or multinominal groups), and 4 (homogeneity of variance).

The one-way ANOVA is a robust test, such that some violation of the normality assumption may be tolerated, such that the test will still provide valid results; you can accept an approximately normal distribution.

The results of Levene's test (to test assumption 4) are presented in the first section of the one-way ANOVA table in SPSS output tables (as discussed further in the tutorial). If the population variance of both groups is equal, this test returns $p > 0.05$, indicating that you have met the assumption of homogeneity of variances. You should then report the results in SPSS for "equal variances assumed," meaning that you report the results of the standard one-way ANOVA and will use the Tukey post hoc test. However, if the test returns $p < 0.05$, the population variances are unequal, and you have violated the assumption of homogeneity of variances. Therefore, you should report the results in SPSS for "equal variances not assumed," meaning that you report the results of the one-way Welch's test ANOVA (instead of the one-way ANOVA), and you will use the Games-Howell post hoc test (instead of the Tukey post hoc test).

Reporting – Begin by reporting that you met the outliers and normality assumptions (assumptions 1 and 2, respectively), and state the value of

Levene's test (assumption 4). In SPSS, one of the output tables of the test will contain the descriptive statistics. Begin by reporting the mean and SD (mean ± SD) for your groups. The rest will depend on the results of Levene's test as given here:

- *If Levene's test yields* p > *0.05:* You will report the results of the standard one-way ANOVA. If there is a significant p-value, you should report that there is a significant difference between the groups. Now, to determine which groups are different, you should use a Tukey post hoc test. You should report its results by stating which groups are significantly different from each other and the p-values for these differences.
- *If Levene's test yields* p < *0.05:* You will report the results of the one-way Welch's ANOVA. If there is a significant p-value, you should report that there is a significant difference between the groups. Now, to determine which groups are different, you should use a Games-Howell post-hoc test. You should report its results by stating which groups are significantly different from each other and the p-values for these differences.

Kruskal-Wallis H test – The Kruskal-Wallis H test can be thought of as the nonparametric version of the one-way ANOVA test, and is also called *one-way ANOVA on ranks*. You may consider the one-way ANOVA when you want to find if there is a significant difference between a continuous dependent variable and an ordinal (or multinominal) independent variable and your data do not meet the normality assumption of the one-way ANOVA, or when your dependent variable is ordinal (but you can use it with continuous dependent variable). As in a one-way ANOVA, if you are interested in knowing which groups are statistically significant from each other, you need to run a post hoc analysis using Dunn's procedure with the Bonferroni adjustment. This test analyzes the subgroups when you find that there is a difference between three or more groups, but you want to know exactly between which two subgroups the difference is most significant. By default, SPSS will perform this post hoc analysis if you have a significant Kruskal-Wallis H test, and it will produce the results under "pairwise comparison."

Your null hypothesis ($p > 0.05$) is that the distributions are equal. Your alternative hypothesis will depend on whether your data meet assumption 3 (similar distribution). If your data do meet assumption 3, then your alternative hypothesis ($p < 0.05$) is that the medians of the groups are different. If you fail to meet assumption 3, then your alternative hypothesis ($p < 0.05$) is that the mean ranks of the groups are different.

Assumptions – The assumption of this test is assumption 3 (similar distribution). When you meet this assumption, you may report your results more easily by using medians. But if you violated this assumption, you will need to report your results using the more uncommon concept of mean ranks.

Reporting

- *If reporting medians:* Begin by stating that you found similar distributions using a box plot. Add a descriptive table that includes medians for your groups. Then state the difference in medians, followed by the p-value. If you have a significant p-value, report which groups were significantly different from each other (via post hoc analysis), and the p-values of these differences.
- *If reporting mean ranks:* Begin by stating that you found different distributions using a box plot. Report the mean ranks of the groups and whether the difference is significant, followed by the p-value. If there was a significant p-value, report which groups were significantly different from each other (via post hoc analysis), and the p-values of these differences.

Chi-square and fischer's exact tests – The chi-square test and Fischer's exact test are the most popular tests. They both compare two variables (dependent and independent) in a contingency table (also called a *cross-tabulation table*), which shows the distribution of one variable in rows and another in columns) to see if they are related. They tell you how much of a difference there is between your observed counts and the counts you would have expected if there were no relationship at all between the variables. There are several variations of the chi-square statistic, as you might use it to find the differences or the association between two variables, the choice of which depends upon how you collected the data and which hypothesis is being tested. However, all of the variations use the same idea—namely, that you are comparing expected values with the values you actually collect.

Fischer's exact test is also used to analyze contingency tables. Whereas a chi-square test can be used in more than 2 × 2 tables (where you can have multinominal variables), it is not the case in Fischer's exact test, where you should only have dichotomous variables. On the other hand, Fischer's exact test can be used when your data fail the assumption of sample size (assumption 5), an assumption required for the chi-square test.

Your null hypothesis ($p > 0.05$) is that the proportion of your dependent variable equals the proportion of your independent variable. Your alternative hypothesis ($p < 0.05$) is that the proportions are not equal. If you get a statistically significant result for the chi-square test, you can establish where the differences are by running a post hoc test called the *z test of proportions*. This test helps you to determine which of the two groups of your independent variable differ in terms of the three or more categories of the dependent variable. For instance, supposing that you found a significant difference in proportion between working status (yes or no) and country (Jordan, the United States, and the United Kingdom), and you want to find if there is a difference in working status and in which country or countries the working status difference exactly exists, the z test of proportions can answer this question.

Assumption – The main assumption for a chi-square test is assumption 5 (sample size requirement). A general recommendation for the minimum sample size required for the approximation to be sufficient is that no more than 20% of the cells in the contingency table have expected frequencies of 5 or less, and that no cells have expected frequencies less than 1.

Note that if your data did not meet this assumption, SPSS will automatically run Fischer's exact test.

Reporting – Reporting of chi-square and Fischer's exact tests will depend on the contingency tables you produce. Always report the number and percentages as "n(%)"; for example, "53(3%)." Report the significance level and the groups that are significantly different from each other if you ran a post hoc test (z test of proportions).

The z test of proportions is presented as small letters within each cell of the contingency tables, which denote the following:

- If the letters are the *same* (i.e., "a" and "a"), there are *no* statistically significant differences in proportions between the two groups of the independent variable.
- If the letters are *different* (i.e., "a" and "b"), there *are* statistically significant differences in proportions between the two groups of the independent variable.

Reporting on the z test of proportions should follow the chi-square test reporting, and it should include the percentage of proportion for each reported variable. So you can say, "We found a significant difference between working status and country ($p = 0.004$). on the z test of proportion, so people living in the United States are more likely to be working (70%) compared to Jordan (50%)."

Associations – Associations tests are used to find if there are associations among variables (Laerd Statistics, 2017).

Pearson correlation – The Pearson correlation test is used to determine the strength and direction of a linear relationship between two variables on a continuous scale. The test generates a coefficient called the *Pearson correlation coefficient*, denoted as *r*, which is a measure of the strength and direction of a linear relationship between the two variables. Its value can range from -1 for a perfect negative linear relationship (i.e., if one variable increases, the other decreases), to $+1$ for a perfect positive linear relationship (i.e., if one variable increases, the other increases as well). A value of 0 (zero) indicates no relationship between the two variables. Remember that the observations in the continuous variables should be paired such that each observation in one variable can be paired with observations from the other variable.

Your null hypothesis ($p > 0.05$) is that the correlation coefficients are equal to zero, and your alternative hypothesis ($p < 0.05$) is that the correlation coefficients are different.

Assumptions – This test has three main assumptions: assumption 1 (no outliers), assumption 2 (normal distribution, for both variables), and assumption 6 (linear relationship, between the two variables).

Testing for outliers in Pearson's correlation is done by a scatterplot (not a box plot), which is also used to test for a linear relationship.

Reporting – Begin by detailing the results of these assumptions. The first step in interpreting your results is to understand the Pearson's correlation coefficient value, which is a measure of the strength and direction of the association between your two variables. The second step in interpreting your results is to determine whether the Pearson's correlation coefficient value is statistically significant (i.e., $p < 0.05$).

Begin your reporting by describing the variables that you are assessing via descriptive statistics (e.g., age and BMI). Next, state that your variables met the required assumptions (i.e., outliers, distribution, and linearity). Finally, report the p-value of the correlation coefficient for your association testing results.

Spearman correlation – Also known as *Spearman's rank-order correlation*, the Spearman correlation, like the Pearson correlation, measures the strength and direction of a relationship. However, the Spearman correlation coefficient (denoted as ρ, rho, or r_s) is a measure of the association between two continuous variables, two ordinal variables, or a mix of the two. The value of the coefficient is figured the same way as with the Pearson correlation. A value of 0 (zero) indicates no relationship between the two variables. Although Spearman's test does not require a linear relationship between the variables, it does require a monotonic relationship (discussed previously in this chapter). Remember that the observations in the continuous or ordinal variables should be paired such that each observation in one variable can be paired with observations from the other.

Your null hypothesis ($p > 0.05$) is that the correlation coefficients are equal to zero, and your alternative hypothesis ($p < 0.05$) is that the correlation coefficients are different.

Assumptions – The main assumption for this test is assumption 7 (monotonic relationship) between the variables. Violations of this assumption are not correctable; if a violation occurs, you will need to go for another test (i.e., Kendall's tau-b).

Reporting – Begin by explaining this assumption. The first step in interpreting your results is understanding the Spearman's coefficient value (r_s), which is a measure of the strength and direction of the association between your two variables. The second step in interpreting your results is to determine whether the Spearman's rank-order correlation coefficient value is statistically significant (i.e., $p < 0.05$).

Begin your reporting by describing the variables that you are assessing via descriptive statistics (e.g., age and BMI). Then, state that your variables met the required assumption (i.e., monotonic relationship). Finally, report the p-value of the correlation coefficient for your association testing results.

Point-biserial correlation – A point-biserial correlation is a special case of the Pearson correlation that is used to determine the strength of a linear relationship between one continuous variable and one dichotomous variable, rather than two continuous variables. The point-biserial correlation coefficient (denoted as r_{pb}) indicates the strength and direction of the association between the two groups of the dichotomous variable and the continuous variable. The value of the coefficient is figured the same way as with the Pearson correlation. A value of 0 (zero) indicates no relationship between the two variables. Remember that the observations in the continuous or ordinal variables should be paired such that each observation in one variable can be paired with observations from the other variable.

Your null hypothesis ($p > 0.05$) is that the correlation coefficients are equal to zero, and your alternative hypothesis ($p < 0.05$) is that the correlation coefficients are different.

Assumptions – The three main assumptions for this test are assumption 1 (no outliers), 2 (normal distribution), and 4 (equal variances).

Reporting – Begin your reporting by describing the variables that you are assessing via descriptive statistics (e.g., age and BMI). Next, state that your variables met the required assumption (i.e., outliers, distribution, and equal variances). Finally, understand the coefficient value (r_{pb}), which is a measure of the strength and direction of the association between your two variables, and the statistical significance of this relationship via the p-value.

Kendall's tau-b – Kendall's tau-b is a nonparametric alternative to the Pearson's correlation (and sometimes to the Spearman's correlation), which measures the strength direction of a linear relationship between two continuous variables, two ordinal variables, or a mix of the two. The value of the coefficient is figured the same way as with the Pearson correlation. A value of 0 (zero) indicates no relationship between the two variables. Remember that the observations in the continuous or ordinal variables should be paired such that each observation in one variable can be paired with observations from the other variable.

Your null hypothesis ($p > 0.05$) is that the correlation coefficients are equal to zero, and your alternative hypothesis ($p < 0.05$) is that the correlation coefficients are different.

Assumptions – It is preferred to meet assumption 7 (monotonic relationship), but this is not a strict requirement. Therefore, we may consider using Kendall's tau-b as an alternative to the Spearman test when this assumption is not met.

Reporting – Begin your reporting by describing the variables you are assessing via descriptive statistics (e.g., age and BMI). Then, state that your variables met the optional assumption (i.e., monotonic relationship). The first step in interpreting your results is to understand the coefficient value (τ_b), which is a measure of the strength and direction of the association between your two variables. The second step in interpreting your results is to determine whether the correlation coefficient value is statistically significant (i.e., $p < 0.05$).

Tutorial

As discussed in Chapter 2, studying a relationship between two variables in a cross-sectional study design suffers from the disadvantage that you cannot be sure which variable happened first. Therefore, it is difficult to establish a causal relationship with this design. On the other hand, you may be able to establish the presence of a relationship between variables when you are sure that one variable occurred before the other. If so, you may be able to conduct a case-control or a cohort study to further study the relationship between your variables. At this point, we want to remind you to apply for access to a data repository (as detailed in Chapter 2) and obtain a data set so that you end up with a new study as we finish the book.

In the tutorial in Chapter 2, we obtained the data from the Framingham Heart Study, a long-term, prospective cardiovascular cohort study on residents of Framingham, Massachusetts. The study began in 1948 with 5,209 adult subjects (Mahmood et al., 2014). The aim that we proposed for our study and obtained the data for was stated as follows:

> It is well-known that high body mass index (BMI) is one of the risk factors for developing cardiovascular disease. In this study, we will discuss the opposite: the effect of previous cardiovascular disease on BMI.

The data obtained in the previous tutorial comprises 4,434 patients. For each patient, we have data regarding the following (of which we can use all or some of the information): sex (dichotomous data), age (continuous), previous history of cardiovascular events (dichotomous), BMI (continuous), total cholesterol (continuous), systolic and diastolic blood pressure (continuous), glucose (continuous), and smoking history (dichotomous).

Note: The Framingham Heart Study defined previous cardiovascular disease as previous angina pectoris, previous myocardial infarction, or previous stroke. To simplify our discussion in this tutorial, we will consider previous cardiovascular disease as previous myocardial infarction only.

As we stated earlier in this chapter, we always begin by using descriptive statistics (i.e., distribution, central tendency, and dispersion). For each group, descriptive statistics may be generated according to the type of variable.

For nominal and ordinal data, report distribution via frequency tables, choose "analyze," then "Descriptive Statistics," and finally "Frequency." In the box that appears, you can choose the variables you want in order to generate a frequency table similar to Table 3.1, where we generated frequency tables for the dichotomous variables sex, current cigarette smoker, and previous history of cardiovascular disease.

Table 3.1   Frequency Tables for the Dichotomous Variables Sex, Current Cigarette Smoking, and Previous History of Cardiovascular Disease

		Frequency	Percent	Valid Percent	Cumulative Percent
Sex					
Valid	Men	1,944	43.8	43.8	43.8
	Women	2,490	56.2	56.2	100.0
	Total	4,434	100.0	100.0	
Current Cigarette Smoker					
Valid	No	2,253	50.8	50.8	50.8
	Yes	2,181	49.2	49.2	100.0
	Total	4,434	100.0	100.0	
Previous History of Cardiovascular Disease					
Valid	No	4,348	98.1	98.1	98.1
	Yes	86	1.9	1.9	100.0
	Total	4,434	100.0	100.0	

Note: We usually look for the frequency and the percent (the second and third columns) in our data reporting.

For continuous data→report mean and SD via descriptive tables, by choose "Analyze," then "Descriptive Statics," and finally "Descriptive." In the box that appears, you can choose the variables you want in order to generate a descriptive table similar to Table 3.2, where we generated a descriptive table for the continuous variables age, BMI, serum cholesterol, systolic blood pressure, diastolic blood pressure, and cigarettes smoked per day.

Then we will do the inferential statistics to test our hypothesis, which we will formulate from our aim:

- *The null hypothesis*: There is no difference in BMI in patients with a history of previous cardiovascular disease compared to those without.
- *The alternative hypothesis*: There is a difference in BMI in patients with a history of previous cardiovascular disease compared to those without.

We will analyze the relationship between BMI and previous history of cardiovascular disease as given here.

Step 1: Determine your study groups—Based on the data we have, we will do a cross-sectional study with between subjects analysis (i.e., patients with previous cardiovascular disease and patients without). We choose not to do a within subjects analysis because we do not have any longitudinal patient follow-up. Accordingly, we will choose the "Group Differences" column from the graph shown in Figure 3.1.

Table 3.2   Description of the Continuous Variables Age, BMI, Serum Cholesterol, Systolic Blood Pressure, Diastolic Blood Pressure, and Cigarettes per Day

	N	Minimum	Maximum	Mean	SD
Age (years)	4,434	32	70	49.93	8.677
BMI (kg/(M²))	4,415	15.54	56.80	25.8462	4.10182
Serum cholesterol (mg/dl)	4,382	107	696	236.98	44.651
Systolic blood pressure (mmHg)	4,434	84	295	132.91	22.422
Diastolic blood pressure (mmHg)	4,434	48	143	83.08	12.056
Serum glucose (mg/dl)	4,037	40	394	82.19	24.400
Cigarettes per day	4,402	0	70	8.97	11.932

Step 2: Determine the number and types of study variables—For our hypothesis, there are at least two variables (namely, BMI and history of cardiovascular disease). BMI is a continuous variable, whereas previous history of cardiovascular disease is a dichotomous variable (i.e., the result is yes or no). There are also other variables that may be used in this study to support our theory (i.e., smoking status, serum cholesterol, etc.)

You should also draw a distinction between the study variables as to whether they are independent or dependent. Previous history of cardiovascular disease is an independent variable, while BMI is a dependent variable.

Accordingly, and as shown in Figure 3.3, as the dependent variable is continuous and the independent variable is dichotomous, we have narrowed down our options to the independent samples t test and Mann-Whitney U test. We will choose which of these tests to use depending on our third step (i.e., assumption testing).

Step 3: Assumption testing: To choose between the two main tests for each of the independent variables, we need first to test both for outliers (via box plot, assumption 1) and normality (via Shapiro-Wilk test, assumption 2) in the SPSS program, and we will test for the fourth assumption (i.e., homogeneity of variance) afterward. Please consult the SPSS guide on how to test for these on SPSS according to the SPSS version you have (SPSS citation). Figure 3.4 shows the output screen for these tests.

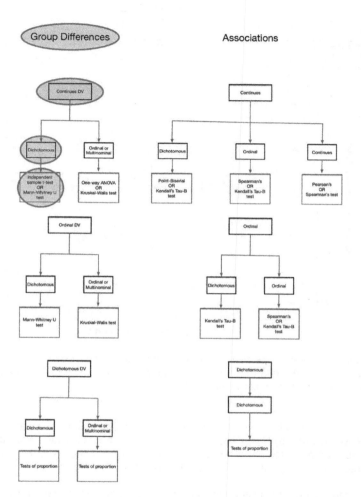

Figure 3.3   Example of how to choose the most appropriate statistical test (steps 1 and 2).

In these results, we have several outliers (>1.5 plot lengths but <3 box lengths) and extreme outliers (>3 box lengths). This means that we do not meet assumption 1. In addition, the Shapiro-Wilk test results are significant (<0.05) for patients without a previous history of cardiovascular disease, although it wasn't significant for patients with a previous history of cardiovascular disease. These results mean that we do not meet assumption 2. We do not need to test for assumption 4 (i.e., homogeneity of variance), as we won't go for a parametric test, due to the fact that our data violated assumptions 1 and 2.

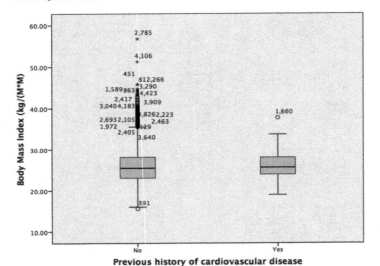

Tests of Normality

	previous history of cardiovascular disease	Kolmogorov-Smirnova			Shapiro-Wilk		
		Statistic	Degree of freedom	Sig.	Statistic	Degree of freedom	Sig.
Body Mass Index (kg/(M*M)	No	.053	4329	.000	.957	4329	.000
	Yes	.084	86	.197	.976	86	.108

a. Lilliefors Significance Correction

Figure 3.4   Results of tests for normality (Shapiro-Wilk test) and outliers (box plot) for the relationship between BMI and the independent variables of previous history of cardiovascular disease.

Therefore, we will use the Mann-Whitney U test to evaluate our data. For the Mann-Whitney U test, we now just need to test for a similar distribution (assumption 3), so that we know how to report our results. The output screen for assumption 3 is shown in Figure 3.5.

We can see from the histograms that the distributions are similar for BMI for patients with and without a previous history of cardiovascular disease, so we will need to use medians when reporting our results (although it is not symmetrical, the present similarity is enough to use medians). The results of the Mann-Whitney U test are shown in Figure 3.6.

The test results in the manuscript can be reported as follows:

A Mann-Whitney U test was run to determine if there were differences in BMI between patients with a previous history of cardiovascular disease and those without. Distributions of the engagement scores for males and females were similar, as

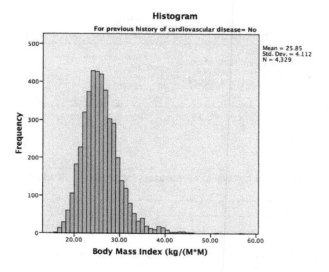

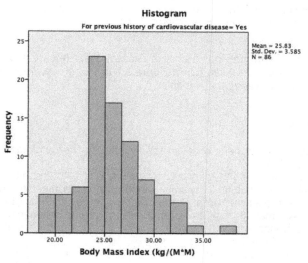

Figure 3.5   Distribution testing between BMI and the independent variable of previous history of cardiovascular disease.

assessed by visual inspection of histograms. Median BMI was not statistically significant ($p = 0.716$), as the median in patients with a previous history of cardiovascular disease was 25.56, compared to 25.45 for patients without a previous history of cardiovascular disease.

Hypothesis Test Summary

Null Hypothesis	Test	Sig.	Decision
1	The distribution of Body Mass Index (kg/(M*M) is the same across categories of Prevalent MI (Hosp,Silent).	Independent-Samples Mann-Whitney U Test	.716	Retain the null hypothesis.

Asymptotic significances are displayed. The significance level is .05.

Previous history of cardiovascular disease	Median
Yes	25.56
No	25.45

Figure 3.6   Mann-Whitney test results for BMI and the independent variable of previous history of cardiovascular disease. Note that the significance level obtained is 0.716 ($p \geq 0.05$), so the SPSS automatically displayed the message "Retain the null hypothesis."

What we did with previous history of cardiovascular disease and BMI was just one example. You can follow Figure 3.1 to test group differences and associations between other variables.

References

Bigby M. Odds ratios and relative risks. *Archives of Dermatology*. 2000 Jun 1;136(6):770–772.

Button KS, Ioannidis JP, Mokrysz C, Nosek BA, Flint J, Robinson ES, Munafò MR. Power failure: Why small sample size undermines the reliability of neuroscience. *Nature Reviews Neuroscience*. 2013 May;14(5):365.

Conover WJ, Iman RL. Rank transformations as a bridge between parametric and nonparametric statistics. *The American Statistician*. 1981 Aug 1;35(3):124–129.

Cummiskey K, Kuiper S, Sturdivant R. Using classroom data to teach students about data cleaning and testing assumptions. *Frontiers in Psychology*. 2012 Sep 25;3:354.

Esar E. *Dictionary of humorous quotations*. 1953.

Faul F, Erdfelder E, Buchner A, Lang AG. Statistical power analyses using G* Power 3.1: Tests for correlation and regression analyses. *Behavior Research Methods*. 2009 Nov 1;41(4):1149–1160.

Garson GD. *Testing statistical assumptions*. Asheboro, NC: Statistical Associates Publishing, 2012.

Gonzalez-Mayo A, James TA, (Firm) KM. *USMLE Step 2 CK lecture notes 2016: Psychiatry, epidemiology, ethics, patient safety*. 2015.

IBM Corp. Released 2012. IBM SPSS Statistics for Macintosh, Version 21.0. Armonk, NY: IBM Corp.

Kothari CR. *Research methodology: Methods and techniques*. New Age International; 2004.

Laerd Statistics. Statistical tutorials and software guides, 2017. Retrieved in January 2017, from https://statistics.laerd.com/.

Le CT, Eberly LE. *Introductory biostatistics*. John Wiley & Sons, 2016.

Mahmood SS, Levy D, Vasan RS, Wang TJ. The Framingham Heart Study and the epidemiology of cardiovascular disease: A historical perspective. *Lancet*. 2014 Mar 15;383(9921):999–1008.

Marshall E, Boggis E. *The statistics tutor's quick guide to commonly used statistical tests*. University of Sheffield, England. Available from http://www.statstutor.ac.uk/resources/uploaded/tutorsquickguidetostatistics.pdf. 2016.

Nayak BK, Hazra A. How to choose the right statistical test? *Indian Journal of Ophthalmology*. 2011 Mar;59(2):85.

Nuzzo R. Scientific method: Statistical errors. *Nature News*. 2014 Feb 13;506(7487):150.

Peat J, Barton B. *Medical statistics: A guide to data analysis and critical appraisal*. John Wiley & Sons, 2008.

Wasserstein RL, Lazar NA. The ASA's statement on p-values: Context, process, and purpose. *The American Statistician*, 2016;70(2):129–133, DOI: 10.1080/00031305.2016.1154108.

4

From Research Project to Article

Saif Aldeen Saleh AlRyalat and Shaher Momani

Writing the manuscript

At this point, you have completed most of the work needed for your research project; now you just need to report your research project as a manuscript. The manuscript is your means of communicating your research findings to others, so that others may benefit from it. Your project may contribute to systematic reviews, influence clinical guidelines, and change clinical practices. It is very important to master manuscript writing because readers would question the reliability of even the most sophisticated and valid research if it is poorly communicated.

Over the past several decades, a vast body of literature failed to adhere to the basic principles of research reporting, one of which is to include all information about methods and results (Altman and Moher, 2014). An international initiative called the EQUATOR Network (where the name stands for "Enhancing the Quality and Transparency of Health Research") is a coordinated attempt to improve the reliability and value of published health research literature by promoting transparent and accurate reporting and wider use of robust reporting guidelines (Equator Network 2009). Table 4.1 provides an overview of the reporting guidelines available through EQUATOR. As we are concerned with observational studies in this book, we will rely on the Strengthening the Reporting of Observational Studies in Epidemiology (STROBE) statement, which is a set of recommendations for improving the reporting of observational studies. However, research on open data requires further recommendations that weren't addressed in the STROBE statement. Thus, an initiative called "The REporting of studies Conducted using Observational Routinely collected Data (RECORD)" was established as an expansion of the STROBE statement to address specific reporting issues relevant to research using routinely collected health data, which is related to research on open data (Benchimol et al., 2015). This section is highly dependent on these reporting guidelines, as most scientific journals require adherence to these guidelines for submission, so this chapter will cover most of these recommendations, in addition to some specifics related to open data research.

Remember that every author has his or her own writing style, and you will eventually adopt your own. Here, however, we propose a style recommended by several experts as the best one to follow for writing a manuscript derived from open data. We recommend beginning with writing the "Methods"

Table 4.1   Overview of the Reporting Guidelines Available Through the EQUATOR Network

Initiative	Study Design	Source
CONSORT	Randomized controlled trials	http://www.consort-statement.org
PRISMA	Systematic reviews and meta-analysis	http://prisma-statement.org
STROBE	Observational studies	https://www.strobe-statement.org
CARE	Case reports	http://www.care-statement.org

section (which might also have "Methodology" or "Materials and Methods" as the title), and dividing it into four subsections; study design, participants and settings, variables, and statistical analysis. After that, report your results in a special section with that title, beginning by describing results for the sample included, and then giving the main results of your study. Then move on and write a "Discussion" section, beginning with a summary paragraph, several paragraphs to interpret the results, and a paragraph describing the limitations of your experiment. Then go back and write an "Introduction" section before the "Methods" section, with a length of one to one-and-a-half pages, as described later in this chapter. Finally, write any supplementary sections. It is extremely important to plan out the Discussion and Introduction before writing them, as if you are putting a title for each paragraph, so that you keep each paragraph focused and you won't repeat yourself unnecessarily.

In this chapter, we will walk you through the details of writing the Introduction, Methods, Results, and Discussion (IMRD) sections of your manuscript. After that, we will discuss the Abstract, title page, References, and Acknowledgments sections (Von Elm et al., 2007; American Medical Association, 2009). At the end of this chapter, a tutorial that summarizes writing principles will remind you with the specifics when you write your manuscript. We recommend keeping it as a checklist to ensure that your manuscript adheres fully to reporting guidelines (whether they are the general ones, or ones stipulated by the journal that you want to submit your paper to for publication). Moreover, the referenced annotated article will also aid you during manuscript writing (Pettis et al., 2013).

The four main sections of a manuscript

Methods – The "Methods" section is devoted to explaining how you reached the results that you will report in the "Results" section. The "Methods" section in open data research are usually derived from either the original article's methodology, where it already described what the researchers did in their experiment, or the protocol associated with the data in the repository. You need to refer to it before writing about your methods. Divide the coverage of your methods into the following sections:

Design – Begin by specifically naming your study design, followed by an explanation of that design (refer to Chapter 2) (e.g., "We used a cross-sectional design in which we evaluated our patients in the period January–June 2017."). This presentation of your study design allows professionals and nonprofessionals alike to understand it, as you are using universal terminology to give your explanations. Key explanations that you should incorporate with each study design include the following:

- *Cohort:* Your sample population, duration of follow-up, and the exposure
- *Case-control:* The criteria for choosing the cases and their differences from the control, and the population from which you chose the participants.
- *Cross-sectional:* The population and point of time that your cross section took place

During the explanation, try not to use the terms *prospective* or *retrospective,* as some readers may perceive the words *cohort* and *prospective* as synonymous, and may reserve the word *retrospective* for a case-control study or retrospective cohort study. Therefore, instead of using these words, you should describe in detail how and when data collection took place. When you are using open data, you need to describe how the original study collected data, as if you were one of the original data collectors. For example, do not just say, "We obtained data from their repository." Remember that the secondary use of existing data is a creative part of observational research and does not necessarily make the results less credible or important. However, briefly restating the original study's aims might help readers understand the context of the research and possible limitations of the data.

Setting and participants – Readers need information about the setting, including the location and duration of the original study, which usually reflect that information about the data collection. Provide details about the location and recruitment site, including the country, city, and institution names when possible. When discussing the duration, specify dates rather than describing the length of time.

For participants, it is important to differentiate between the original data-source population and the population that you included in your study. They might not be the same, as in the following scenarios:

- You only included part of the original data, which might be due to miss-ing data in the rest or any other reason. In this case, you need to detail the inclusion criteria for the sample population you included from the original study.
- You combine data from two or more sources of open data. In this case, you need to state explicitly that your study's sample population is a *Linkage* from two or more data sets. Here, it is preferable to dem-onstrate, using flowcharts or other diagrams, how your study's sample population is related to the original populations.

A description of the participants helps readers to understand the applicability of the results. For example, in our study discussing the effect of previous history of cardiovascular disease on a patient's body mass index (BMI), describing that your included sample is from elderly patients will help your readers interpret the findings in the context of the included sample age, so they won't generalize the results on young age groups. Describe how you chose the number of participants (or the sample size). The most essential part of describing the participants is to describe the eligibility criteria for your study participants and how you chose them (i.e., inclusion and exclusion criteria).

Another part of your description should depend on the study design, as given here:

- For a cohort study, detail the type of exposure (e.g., smoking) you studied, how you followed your participants, and how you handled loss of follow-up (i.e., when some of the patients included at baseline won't participate in subsequent follow-ups).
- For a case-control study, detail the definitive criteria you used to define the participant as a "case," and how you chose a matching control. *Matching* is finding a participant that is similar to the case in all characteristics, but different in the particular characteristic that defines the participant as a case. To simplify this concept, say that we want to study environmental risk factors for developing a heart attack. We would define a "case" as a participant who had a heart attack. A participant who did not have a heart attack would be considered a "control." We would match the controls with the cases by characteristics such as age and gender (i.e., we would match participants who had a heart attack with patients who did not have a heart attack at a similar age and gender). This would eliminate characteristics (such as age and gender) contributing as risk factors for a heart attack, which will allow us to study environmental factors precisely.
- For a cross-sectional study, detail the source and method for selecting participants. Also, include whether you applied a sampling strategy (e.g., simple random sampling).

Variables – Authors should define all variables, including outcomes, exposures, predictors (independent variables), potential confounders, and potential effect modifiers (i.e., confounding variables). Provide sources of data and details of your methods of assessment (i.e., the methods you used to measure the variables). Disease outcomes require adequately detailed descriptions of the diagnostic criteria, so when your outcome measure is a disease like myocardial infarction (MI), you need to diagnose MI according to the most recent guidelines.

The authors would need to provide detailed descriptions of the criteria defining a case in a case-control study, a disease event during follow-up in a cohort study, and the prevalent disease in a cross-sectional study.

Statistical analysis – Begin by describing the statistical software you used (e.g., SPSS). Describe any manipulation you did to your data. For example, you may have grouped a continuous variable (e.g., age) to create a new categorical variable (e.g., age groups). As a guiding principle, statistical methods should be described with enough detail to enable a knowledgeable reader with access to the original data to verify the reported results. You should also describe the extent to which you had access to the original open data used to create your study's population.

Results – The "Results" section is the reason you did your research in the first place. It should provide only a factual account of the data you found; interpretations and descriptive texts reflecting your views or opinions should not be included in this section.

Descriptive data – In this part, you should provide a description of your sample with sufficient detail to allow readers to judge its generalizability, as detailed later in this chapter. Begin by providing the number of participants in your study. Give the initial number you intended to include, followed by the final number actually in the study, with details of excluded participants and the reasons for their exclusion. This will help readers judge whether the study population was representative of the target population and whether bias was possibly introduced. Especially in a cohort study, detail the numbers of participants lost to follow-up for each variable. An informative and well-structured flow diagram may readily and transparently convey information about inclusion criteria, exclusion criteria, and loss to follow-up, which might require a lengthy description otherwise (see Egger et al., 2001).

For descriptive data purposes, we have two types of variables: continuous and categorical. With continuous variables, you should report the mean and standard deviation (or median asymmetrical distribution). With categorical variables (both nominal and ordinal), you should report numbers and proportions. Inferential measures such as standard errors, confidence intervals, and significance tests should be avoided in descriptive tables, and should be included in inferential statistics. It is preferable to list descriptive data in tables to explain the results more visually and concisely.

Primary outcome – After providing the reader with a sufficient description of your data, you may begin presenting your main results. By *primary outcome,* we mean the results related to your primary objective. Begin with your most important findings. The authors should explain all potential confounders, and the criteria for excluding or including certain variables in statistical models. An example for a primary outcome is the result of the tutorial in Chapter 3, on the relation between previous history of cardiovascular disease and BMI.

Secondary outcome (optional) – In addition to the main analysis, other analyses are often done in observational studies, and they represent your secondary outcome. They may address specific subgroups, the potential interactions among risk factors, or alternative definitions of study variables. An example from our previous tutorial would be the results of analyzing the relation between smoking and BMI.

Tables and figures – It might be more understandable to readers if some results are presented in a figure, or if page-long results are summarized in a simple table. Tables and figures are tools you can (and should) use to deliver your ideas in clearer language. Remember that you should not replicate your results in both forms (e.g., both written and in a figure), although you can focus on some of the important results presented in a table in the text.

Discussion – The "Discussion" section addresses the meaning of the study. Here, you can translate the rigid results of your study into a more interpretable answer to your research question. Almost universal parts of this section are described in the following subsections.

Key results – Begin by summarizing your main results with reference to your study's objectives. A short summary reminds readers of the main findings and may help them assess whether the subsequent interpretation and implications offered by the authors are supported by their findings. Remember that, unlike the "Results" section, you may freely cite other references here.

Interpretation – This is the heart of your study, where you present your interpretation of the results with references to other studies, for the purpose of comparing your results with others. The most important thing to remember here is that you must not overinterpret your results. Overinterpretation is common and human: Even when we try hard to give an objective assessment, we often go too far in some respects. Always keep your study design in mind; it is almost impossible to establish causality in a cross-sectional study, so do not claim to do so if you are using this design. Try not to overestimate a relationship between two variables in a cross-sectional study by claiming cause and effect. A great discussion on how to think about causality is presented by Bradford Hill (1965).

Generalizability – After discussing and interpreting your results, you may need to generalize your results to populations that differ from those enrolled in the study. This part is dependent upon your explanation in the "Methods" section about study design, setting, and participants. For example, you might investigate the average height of a city's population, but to the extent that this city has people representing a whole country, and your sample is representative of the country (as described in the methodology), you might generalize your results accordingly. Thus, it is crucial that researchers provide readers with adequate information about their methods. Authors should compare their study with other preexisting studies. In this way, each study makes its own contribution to the literature, not acting as a stand-alone work for inference and action.

Limitations – The identification and discussion of the limitations of a study are essential parts of scientific reporting. There is no "perfect" study; even those published in high-impact journals have their limitations. But it is important that you report your study's limitations, instead of waiting for reviewers to uncover and report them. When discussing limitations of open data research, discuss the implications of using open data that were not created or collected originally to answer the research question you had. Include in the discussion missing data, unmeasured confounding variables, and other limitations of the original data reported in the original primary publications (if present).

Introduction – Now that you have a full picture of your study, you need to write an "Introduction" section that describes why the study was done and what questions and hypotheses it addresses. The "Introduction" should allow others to understand the study's context and judge its potential contribution to current knowledge. The text should have a funnel-like flow, beginning with general background information and ending with more specific background information about your topic of interest. In the last paragraph, you should state your objectives.

Background – The scientific background of the study provides important context for readers. It sets the stage for the study and describes its focus. It gives an overview of what is already known about a topic and what gaps in current knowledge will be addressed by the study. Background material should include recent, pertinent studies, including any pertinent systematic reviews.

Objectives – Here, you should state your research question or hypothesis. You should specify the populations, exposures, outcomes, and parameters that will be estimated. You should use similar wording here as with the first paragraph of the "Discussion" section, where you showed that your study objective was completed.

Complementary sections of the manuscript

The rest of the sections of your paper will mainly derive from the four initial sections; they include the title, abstract, conclusion, list of references, authors, acknowledgments, and ethical statement.

Title and abstract – We previously described how your research questions should be fully descriptive of your study. This also applies to the title. Readers should be able to easily identify the study design that was used from your title. Including an explicit, commonly used term for the study design (e.g., cross-sectional study) in your title also helps to ensure the correct indexing of articles in electronic databases. Try not to use questions as a title, but use their answers.

The abstract provides key information that enables readers to understand a study and decide whether to read the article. It is important to point out clearly that your study is done with open data, and it is preferable either to mention the name of this data or give a link for it. Abstracts should present only information that is provided in the article, and, due to the limited words you have (as required by most scientific journals), you should only present the most important and interesting information. Typical components include brief summaries of the background, methods, results, and conclusion.

The Results and Conclusion are the main parts you should focus on, presenting key results in numerical form, including the number of participants and p-values. It is insufficient to only state that an exposure is or is not significantly associated with an outcome. The "methods" part of the abstract is the most suitable for stating that your study was done using open data and giving the name of or the link to the data used. Although some journals have more generous guidelines for their abstracts, it's a good idea to keep yours to a maximum of 250 words to ensure its compatibility with most journals, as it is sometimes quite difficult to cut down on the abstract after it has been written.

Keywords – *Keywords* are words that databases use to categorize your study topic. It is important to choose keywords that help the most appropriate readers find your study, and, as a result, provide you with new citations. Spend some time thinking about the best keywords for your study, and try to use vocabulary databases like medical subject headings (MeSHes) (see Chapter 1), which are used by PubMed to index articles.

Conclusion – The "Conclusion" section is the final chance for you to convince readers that your paper is valid. There are two things to remember here. First, you should not repeat what you had already stated in the abstract or in the first paragraph of the "Discussion" section. Second, you should make a great effort to state your results and their importance in one paragraph in a way that will catch the interest of the readers.

References – While writing the paper, cite the references you used after each corresponding sentence or phrase using the first author's name and year of publication (e.g., AlRyalat, 2017), and maintain a record of that article. After finishing your paper, create the list of references, making sure not to add or remove anything from your paper that would cause mistakes in referencing (e.g., referencing the wrong article). You can use Google Scholar to cite the reference in different styles. Using an author's last name, year of publications, and a keyword for your topic, you can search Google Scholar directly. Your reference should be the top result, with different styles provided using citing icon (") beneath each reference (e.g., Vancouver, APA, ...etc.).

It is worth mentioning that there are a number of useful software programs that may help you manage the references you use. Endnote is currently the most popular software for this purpose. It is a subscription-only reference-management software that helps you to save your references and lists them in the style required by journals.

Authors and acknowledgments – On the title page, you should mention the data repository and any supplementary material related to the data and relevant to your study. Describe how others can access these data. List all authors who contributed significantly to the work (see Chapter 1), and acknowledge those who contributed but did not reach the degree of authorship (e.g., the person who only performed statistical analysis).

Ethical Statement – You should disclose all conflicts of interest that you or any of the authors might have in a special section. Disclose any funds received, as your or your fellow authors may have conflicts of interest that may influence aspects of the study. The role of funders also should be described in detail, including the parts of the study for which they took direct responsibility (e.g., design, data collection, analysis, drafting of the manuscript, decision to publish). Refer to Chapter 1 for a detailed description of the ethical aspect of research.

Cover Letter – Now, you've completed your manuscript, and it is now ready for submission. During the submission process, you will be asked to provide a cover letter, which should be written with the same care as the manuscript (some editors judge the integrity of the whole study based on the cover letter). It is a good approach to adopt a consistent format of a cover letter to be used in all your submissions. The following is a suggested format, where you just need to fill the gaps based on your background and your study's information. With time, you will make your own touches on the cover letter format.

[Your name, affiliation, and contact address]

[Date]

Dear [editor name]:

We are glad to submit our manuscript entitled:

[The title of your manuscript]

[Add a paragraph describing your manuscript briefly and professionally]

This manuscript has not been published and is not under consideration for publication elsewhere. We have no conflicts of interest to disclose [*Or mention any that you have*].

Thank you for your consideration!

Sincerely,

[Your name and degree]

Corresponding author.

Publishing the manuscript

Choosing the best journal

Now that you have your manuscript ready for submission, and you already know the life cycle of a manuscript and where it will go after submission (from Chapter 2), how do you choose where to submit? There are three main factors you should consider before deciding to which journal to submit the paper. First, the scope of your manuscript should match the scope of the journal (obviously, you should not submit your medical study to an agricultural journal), so you need to check the scope of any journal you want to submit to, to see that it is suitable. Second, the strength of the journal is another important consideration. It is an easy task to find a journal with matching scope, but to determine the strength of a journal you need to be familiar with its citation index (discussed later in this chapter). Third, if you want your article to be available to readers for free (in which case you need to pay the journal), or you are not willing to pay to the journal, so the reader will have to pay to read your article.

Scope – If you have not already decided where to publish, the most straightforward way to do so is to look for a suitable journal in which your references published. As you were doing your literature review and manuscript writing, you already had encountered studies that had a similar scope to yours. Look at where these studies were published and review these journals to see if they are suitable for your paper. Another technique is to simply look for keywords in search engines (e.g., Google), or in indexing websites; however, note that the Institute for Science Information (ISI) and Scopus are subscription-only services).

If you have a preference for some publishers (e.g., open access publishers), you may search for journals on those publishers' websites. Some publishers (e.g., Elsevier and Springer) provide journal finders that help you to search for journals that could be best suited for publishing your article by simply entering your manuscript's title and abstract. Other software (e.g., EndNote) and independent websites (e.g., FindMyJournal) are also available to help you find the best journal for you. Examples are:

Elsevier: www.journalfinder.elsevier.com
Springer: www.journalsuggester.springer.com

Citation index – *Citation* is the act of acknowledging a publication (such as a journal or book) for its use in a published work. The *citation index* uses the number of citations in a journal's articles to measure the strength of the journal. In 1960, the ISI introduced the first citation index for papers published in academic journals, known as the Science Citation Index (SCI), becoming the first indexing service provider (Csiszar, 2017). In the following years, several other indexing services emerged, but most of them still rely on citations analysis as their main technique.

Generally, indexing services evaluate each journal based on its authenticity, the reliability of its peer review process, and other quality measures to check if it is fit for inclusion. After its inclusion in the indexing service, its strength in terms of citation analysis will be calculated. So, if a journal is included in an indexing service, then it is of high quality, but its strength will be determined after analyzing the citations of the article published in it in the last few years. Now, the prestige of any journal is largely determined by how many abstracting and indexing services cover that journal, and which ones those are. Indexing services screen journals for the quality of their editorial boards and the rigorousness of their peer review processes, so they do the job of checking the journal strength for you. In the past few years, it has become a mandatory requirement for the promotion of teaching faculty in medical colleges and institutions to publish in an indexed journal (Dhammi and Haq, 2016). There are two main indexing services that are relied upon by most institutions in scientific fields (Falagas et al., 2008):

Science Citation Index Expanded (SCIE) – SCI was published originally by the ISI and owned by Thompson Reuters (but now it is owned by Clarivate Analytics). It is now the most widely used index for judging the strength of a journal. Each year, SCI publishes a subscription-only report showing the impact of each indexed journal, called the Journal Citation Report (JCR). SCI uses the impact factor (IF), a numerical measure used to compare ISI journals among each other within the same discipline, to determine the impact of a journal. A journal's IF is based on two elements: the numerator and the denominator. The numerator is the number of citations to items published in the journal during the previous two years. The denominator is the number of substantive articles (source items) published in that journal during the same two years. The equation for calculating the IF is shown here, with y representing the current year:

$$IF = (Citations_{y-1} + Citations_{y-2}) / (Publications_{y-1} + Publications_{y-2})$$

Scopus – Launched in 2004 by Elsevier, Scopus indexes the largest number of journals compared to other databases like ISI and PubMed (Falagas et al., 2008). It uses several types of quality measures for each title, including the Hirsch index (h-index), CiteScore, SCImago Journal Rank (SJR), and Source Normalized Impact per Paper (SNIP).

SJR is a relatively new indexing metric that measure the impact of journals as well as countries. It classifies journals into four categories based on their SJR scores, ranging from Q1 (highest score) to Q4 (lowest score). While IF mostly depends on the number of citations, SJR takes into consideration the prestige of the citing journal in its algorithm and excludes self-citations (i.e., an author cites his or her previous research in a new research paper). It is accessed for free, as opposed to the IF of the SCI.

If you want to publish in a journal not indexed in the aforementioned indexing services, then you will have to look into the authenticity and strength of that journal yourself. This is also the case when you receive an email invitation to publish a study in a relatively new journal. You will need to make sure this journal is not simply looking for financial gain through the collection of submission fees for publishing papers. You can check the credentials of a potential journal by examining the following:

- The publisher (most important)
- Its editorial board
- How long the journal has been publishing
- Citations for the published work

You also should run your own search for any previous criticisms of the publisher, as you can almost always find critics of journals who are not authentic.

Open access – In recent years, much criticism has been raised about the high cost of journal subscriptions, which has led even highly prestigious libraries to cut the journal subscriptions that they provide. The emergence of electronic publishing instead of print-based publishing provided a new approach of delivering science, and this led to the concept of open access. Instead of using a model where the reader pays a subscription fee to read an article, the author, the institution, or the funder pays an open access fee so that the reader may read the article for free. This model aimed to share knowledge without cost to the reader—a model that is even more important in developing countries, where journal subscription costs may not be affordable to most readers. In a recent analysis, the frequency of open access publishing in Jordan increased dramatically in the last 10 years from 7.3% in 2008 to 18.9% in 2018 (AlRyalat and Malkawi, 2018). Publishing in open access model should be encouraged through discounts from publishers and funds from countries and institutions to increase the dissemination of knowledge without cost.

Since the introduction of the open access model of publishing in 2002, several publishers have taken advantage of this model by running journals without authentic peer review, and they have accepted any incoming manuscript for the sake of receiving payment (Beall, 2012). The term *predatory* has been used to refer to these journals to denote this goal of getting money (Beall, 2012). These journals might state the phrase *peer review* in their descriptions, but they do not actually conduct a peer review. You have to check each journal as we stated before earlier in this chapter, before submission.

We may classify journals into three types, based on its open access policy (Harnad, 2004):

- Open-access journals (also known as *gold open access*)
- Subscription-only journals
- Hybrid journals

Subscription-only journals restrict access to paying readers, and the author can't even choose to pay for an open access option. With hybrid journals, the author may choose either model for submission. Gold open access journals sometimes require authors to make a payment in order to make their papers accessible freely to readers. Note that hybrid journals generally have higher costs for open access than open access–only journals; the average payment is $1,865 per article published in an open access journal, compared to an average $2,887 is paid per article published in a hybrid journal (Al-Khatib and da Silva, 2017). More information about how to choose a journal to submit to may be found in Chapter 1.

Self-archiving

Suppose that you do not have the required fee to publish in an open access journal, but you still want to show the results of your research to your peers, especially those who work in your institution. With the option of *self-archiving,* you can publish your research in a non–open access manuscript. Self-archiving is often used as a synonym for *green open access,* where authors add their articles, mostly in the form of presubmission or prepeer review manuscripts that preceded the finalized version, to a repository freely available on the Web (mostly their institutional repository) (Björk, 2014). This is beneficial to both authors and readers in the research field. For authors, providing an open access, freely available version will bring more citations (Wagner, 2010). For readers, they will have a manuscript that is close enough to the published article—at least sufficient for prepurchase evaluation, if not direct citation. The key purpose of green open access is to increase the dissemination of your research results by making them available and citable, especially to nonsubscribing authors, and to those with limited resources to subscribe to the journal or buy the published version of the article.

Publishers usually have their own regulations for self-archiving, and each allows deposition of different manuscript stages. Figure 4.1 shows various manuscript stages and their common description by publishers.

Two main types of repositories on which you can archive your research include the following:

- Individual repository—The easiest way to create an individual repository would be using ResearchGate (researchgate.net).
- Institutional repository—You can check if your institution has its own repository on the Directory of Open Access Repositories (http://www.opendoar.org), or the Registry of Open Access Repositories (http://roar.eprints.org).

Web search engines (e.g., Google Scholar) can find the openly accessible version of the article, especially if this version's title is still the same as the published version.

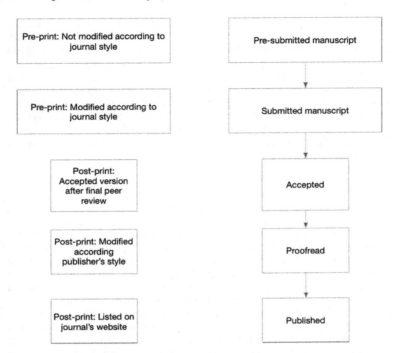

Figure 4.1　The stages any study's manuscript will go through, from presubmission to publishing, and its corresponding description by publishers. You have to check the website of the specific publisher that you wish to contact to check which stage or stages of a paper that you are allowed to deposit freely. Note that most publishers at least allow the deposition of the presubmitted version.

English editing services

Especially if English is not your first language, it is a good idea to have a professional examine and correct your article for any language-related errors. To achieve this, we have two options from which to choose:

- *Editing services*: The editor reviews and changes the text with the intent to improve the flow and overall quality of the manuscript. The focus is on making changes that make an article easier to understand, better organized, and more suitable for the audience. An editor has the freedom to remove entire sentences or rewrite entire paragraphs. You also may specify the journal that you would like to submit to, so that the editor may adapt the manuscript to that journal's style.
- *Proofreading services*: This process involves examining the final draft of the manuscript to ensure that there are absolutely no errors. It focuses on correcting superficial errors; therefore, it normally occurs at the end

of the writing process, as a final step prior to submitting a paper that otherwise is ready to be published. Editing usually includes proofreading. Almost all journals have a proofreading step after article acceptance, and the article is usually double-checked by the corresponding author.

It is a good investment to have your manuscript undergo English-language editing before submission. The cost of these services will depend on the length of your manuscript and the time frame you have for completion.

Response to the publication decision

Receiving an acceptance or a rejection is final, as discussed in Chapter 2. However, there is another option: You may receive a request for major or minor changes, which would be an initial step toward your first published article. You would want to take advantage of this valuable opportunity.

Most journals require peer reviewers to give detailed comments—not simply state, "The author should rewrite the Introduction," but say exactly what pages and lines you should modify. You should consider yourself lucky if a reviewer provides detailed feedback; you should show the same enthusiasm and courtesy by responding to all of the reviewer's comments appropriately, either by doing what was asked or by writing a satisfactory reason for why you do not wish to change what was written. However, even if a reviewer's comments are vague, you should still respond with enthusiasm, although admittedly the job in this case might be more difficult. We recommend using the following cover letter as your response form, especially if the journal in question does not have its own response form, so that you won't miss any comment or mislead the editor on any comment. This form details each reviewer you are addressing, the comments provided by that reviewer, your response to these comments, and where you addressed your response in the manuscript (i.e., page and line in the manuscript, if possible).

[Your name, affiliation, and contact address]

[Date]

Dear [editor name]:

Thank you for considering our manuscript entitled:

[The title of your manuscript]

For possible publication in your journal.

Please find the following detailed responses to reviewer comments:

Reviewer	Comment	Response	Changes Done on Page Number and Line Number

Sincerely,

[Your name and degree]

Corresponding author.

Always remember that although a minor or major revision does not automatically mean a commitment for acceptance, even the highest-impact journal will accept your article if you do a complete job in the review process. Therefore, take your time and consult other experts in the field to help you get your paper accepted.

Postpublishing

There is a lot to do after you have gotten your study into print in order to get the most from your publication. In the era of widespread social media, researchers should utilize all the available tools. There are several social media websites designed for researchers to share, comment, and react to articles. Several studies confirmed the importance of postpublishing activities for increasing the reach of a study and increasing the number of citations. Moreover, Altmetrics, a new indexing metric, measures the impact of research on social interaction (i.e., not only citations as the traditional metrics), such as how often research is blogged, tweeted, or even viewed (Williams, 2017).

Some of the most commonly used websites that increase the Altmetric impact and the chance of citations include ResearchGate (researchgate. net), a social networking site founded in 2008 for scientists and researchers to share papers, ask and answer questions, and find collaborators. It is the largest academic social network in terms of active users. You can sign up for it when you have published your first paper. Another website, Kudos (https://www.growkudos.com), takes another approach; it is designed to help authors of research publications to describe their studies in plain language, supplement their descriptions with information that has been created since the initial publication, and follow the impact of these activities on publication metrics. Mendeley (www.mendeley.com), created by Elsevier, takes a similar approach, but with the advantage of having a reference manager that allows you to organize your references in your manuscript and an academic social network that can help you organize your publications, collaborate with others online, and discover the latest research. These are just a few of the social networking websites available to researchers, and more are likely to be created in the future as the needs and desires of researchers evolve.

Research impact for researchers and institutions

Nowadays, the impact of researchers and institutions is measured by the research output and the quality of this output. One of the main benefits of research is capturing funds and awards, as institutions and researchers with more research output and a higher impact are more likely to capture funds and awards (Li and Agha, 2015).

Researchers – Measuring the impact of an author in science and research can be easily done these days, mostly by using postpublishing websites, as discussed earlier in this chapter. For instance, the number of downloads, number of mentions, and number of reads are handy metrics (Mas-Bleda et al., 2014). Still, the high level of citation to an author's work is the major testimony that that person's work has been noted and used by his peers. An annual list of the most highly cited researchers is published by ISI to recognize leading researchers in the sciences and social sciences from around the world. Noteworthy, one of this book's authors (Professor Momani) is one of the highly cited researchers during the most recent years (Clarivate Analytics 2017).

Institutions – Higher education is one of the fastest-growing sectors globally. Institutions of higher education are usually ranked according to different criteria depending on the ranking system, but all ranking systems include research as a measure. The three main international ranking systems are (Bekhradnia, 2016):

- Academic Ranking of World Universities (ARWU), produced by Shanghai Jiao Tong University
- World University Rankings, produced by Times Higher Education (THE)
- QS World University Rankings, produced by Quacquarelli Symonds Ltd (QS)

Table 4.2 shows the weight put on research in each ranking system based on its criteria.

Despite the variability on the weight of research in each ranking system, there is quite a bit of evidence that these systems provide rankings based on research criteria. Although they claim to take account of other dimensions of universities' performance, it is essentially research that they measure (Bekhradnia, 2016).

Table 4.2 The Main Ranking Systems Used Internationally and the Percentage of Weight Put on Research

Ranking System	Initial Year	Total Number of Indicators	Frequency of Publication	Participating Institutions	Total % Focused on Research	Total % Focused on Academics or Teaching Quality
Academic Ranking of World Universities (Shanghai)	2003	6	Annually	500	100	0
Times Higher Education (THE) Supplement	2004	13	Annually	800	65	35
QS World University Ranking	2013	6	Annually	1,000+	20	80

Tutorial

The following tutorial summarizes the first section of this chapter, so it can be used as a quick reminder during manuscript writing. It can be used for all types of observational studies (i.e., cohort, case-control, and cross-sectional), especially for open data research, and it also can be used for teaching purposes.

Title

Describe your study's hypothesis and its design.

Authors

List all the authors who contributed significantly in the study.

Corresponding author

Identify the author responsible for communicating with the journal's and paper's readers during the submission process and afterward.

Acknowledgments

State the data repository and any supplementary material that is related to the data and relevant to your study. Describe how others can access these data. Acknowledge authors who contributed, but not to the degree of authorship.

Abstract

- Should be concise and informative, preferably containing fewer than 250 words (to meet most journals' requirements).
- Should be structured into four sections (Introduction, Methods, Results, Conclusion), and if a journal requires an unstructured abstract, you need to combine the sections.
- Always use the active voice ("We did....," etc.).
- Clearly state your objective or aim at the end of the Introduction section.
- Mention the name of the link for the open data used.
- Due to the word limit in the abstract, try to put most of your content into the "Methods" and "Results" sections.

Keywords
Put at least three keywords, preferably chosen them from the MeSH database on PubMed.

Introduction
It should be funnel-shaped in the following manner:
- Begin this section with broad information about the topic, in the first paragraph.
- In the next paragraph, you should give more specific background about your topic.
- The next paragraphs should include previous studies that discussed ideas that were close to your own.
- Finally, you should mention your aim, in the form of a research question ("In this study, we will answer the following questions") or in the form of a bullet list of objectives ("In this study, we have two main objectives . . .").

Methods
Begin by stating the ethical approvals you obtained for your study.

Design: Name your study design specifically, and begin discussing it. Include the following details in this part according to your study type:

- *Cohort:* Your sample population, the duration of follow-up, and the exposure
- *Case-control:* The criteria that you used to choose the cases and their differences from the control group, and the population you chose from
- *Cross-sectional:* The population and the point of time your cross-section took place

Setting and participants: For the setting, specify the location and duration (specify dates rather than periods of time) of the original data. For participants, detail the population you included for your study from the original population, and focus on the inclusion and exclusion criteria that you used to reach it. Also, depending on the study design, include the following:

- *Cohort:* Detail the type of exposure you are studying (e.g., smoking) and how you are going to follow your participants, and handle loss of follow-up.
- *Case-control:* detail the definitive criteria you that define the participant as "case," and how you are going to choose a matching control.
- *Cross-sectional:* detail the source and methods of selection of participants, and if you apply sampling strategy.

Variables: Define and give a detailed description of all the variables used in your study.

Statistical analysis: Begin by describing the statistical software you will use, and then describe the descriptive statistics you used with your sample, followed by giving the inferential statistics.

Results
Report the statistical analysis results objectively, without interpretations.

Descriptive data: Begin by describing your sample population via the descriptive statistics you used.

Main results: Begin with your most important findings.

Minor results: Mention other results that were not part of your aim, and you might not include them in your discussion.

Tables and figures: Always use tables and figures to make your results clearer and to avoid overly long texts, but do not repeat the text (i.e., you should report your results once, either as text or in tables and figures).

Discussion
The "Discussion" section is where you interpret your results and compare them with those found in the existing literature.

Key results: Begin by summarizing your main results, where you might cite other works (unlike the summary given in the "Conclusion" section).

Interpretation: Compare the results you have with previous studies, and make sure that you include all perspectives about the results (studies with similar and studies with opposing findings).

Generalizability: Discuss how your results might be generalized to other populations.

Limitations: Discuss the implications of using open data that were not created or collected originally to answer the research question you had. Include discussion missing data, unmeasured confounding variables, and other limitations the original data reported in the original primary publications (if present).

Conclusion
You should state your results and their importance in one paragraph that will encourage readers that your point is valid.

Conflicts of interest
Mention all potential conflicts of interest you have regarding your study.

Acknowledgments
Acknowledge those who contributed to the present work.

References
Check your target journal's preferences, although most journals require references in Vancouver style:

Authors, Title, Journal, Year of publication, Volume, Issue, Pages.

References

Al-Khatib A, da Silva JA. Threats to the survival of the author-pays-journal to publish model. *Publishing Research Quarterly.* 2017 Mar 1;33(1):64–70.

AlRyalat SA, Malkawi L. International collaboration and openness in Jordanian research output: A 10-year publications feedback. *Publishing Research Quarterly.* 2018;34:1–10.

Altman DG, Moher D. Importance of transparent reporting of health research. *Guidelines for Reporting Health Research: A User's Manual.* 2014;1–3.

American Medical Association. *AMA manual of style: A guide for authors and editors.* Oxford University Press, 2009.

Beall J. Predatory publishers are corrupting open access. *Nature.* 2012;489:179.

Bekhradnia B. *International university rankings: For good or ill?* Oxford, UK: Higher Education Policy Institute, 2016.

Benchimol EI, Smeeth L, Guttmann A, Harron K, Moher D, Petersen I et al. The REporting of studies Conducted using Observational Routinely-collected health Data (RECORD) statement. *PLoS Medicine.* 2015 Oct 6;12(10):e1001885.

Björk BC, Laakso M, Welling P, Paetau P. Anatomy of green open access. *Journal of the Association for Information Science and Technology.* 2014 Feb 1;65(2):237–250.

Bradford Hill A. The environment and disease: Association or causation? *InProc Royal Soc Med.* 1965;58:295–300.

Clarivate Analytics. Highly Cited Researchers. https://hcr.clarivate.com, 2017. Accessed September 2017.

Csiszar A. The catalogue that made metrics, and changed science. *Nature News.* 2017 Nov 9;551(7679):163.

Dhammi IK, Haq RU. What is indexing? *Indian Journal of Orthopaedics.* 2016 Mar;50(2):115.

Egger M, Jüni P, Bartlett C, Consort Group. Value of flow diagrams in reports of randomized controlled trials. *JAMA.* 2001 Apr 18;285(15):1996–1999.

Equator Network. Enhancing the QUAlity and Transparency Of health Research. 2009. Accessed September 2017; www.equator-network.org.

Falagas ME, Pitsouni EI, Malietzis GA, Pappas G. Comparison of PubMed, Scopus, web of science, and Google scholar: Strengths and weaknesses. *FASEB Journal.* 2008 Feb;22(2):338–342.

Harnad S, Brody T, Vallières FO, Carr L, Hitchcock S, Gingras Y et al. The access/impact problem and the green and gold roads to open access. *Serials Review.* 2004 Jan 1;30(4):310–314.

Li D, Agha L. Big names or big ideas: Do peer-review panels select the best science proposals? *Science.* 2015 Apr 24;348(6233):434–438.

Mas-Bleda A, Thelwall M, Kousha K, Aguillo IF. Do highly cited researchers successfully use the social web? *Scientometrics.* 2014 Oct 1;101(1):337–356.

Pettis JS, Lichtenberg EM, Andree M, Stitzinger J, Rose R. Crop pollination exposes honey bees to pesticides which alters their susceptibility to the gut pathogen Nosema ceranae. *PloS One.* 2013 Jul 24;8(7):e70182.

Von Elm E, Altman DG, Egger M, Pocock SJ, Gøtzsche PC, Vandenbroucke JP, Strobe Initiative. The Strengthening the Reporting of Observational Studies in Epidemiology (STROBE) statement: Guidelines for reporting observational studies. *PLoS Medicine.* 2007 Oct 16;4(10):e296.

Wagner A. Ben. "Open access citation advantage: An annotated bibliography." 2010. http://hdl.handle.net/10477/12951.

Williams AE. Altmetrics: An overview and evaluation. *Online Information Review.* 2017 Jun 12;41(3):311–317.

Index

POCKET GUIDES TO
BIOMEDICAL SCIENCES

https://www.crcpress.com/Pocket-Guides-to-Biomedical-Sciences/
bookseries/
CRCPOCGUITOB

Series Editor
Lijuan Yuan
Virginia Polytechnic Institute and State University

A Guide to AIDS
Omar Bagasra and Donald Gene Pace

Tumors and Cancers: Brain – Central Nervous System
Dongyou Liu

Tumors and Cancers: Head – Neck – Heart – Lung – Gut
Dongyou Liu

A Guide to Bioethics
Emmanuel A. Kornyo

Tumors and Cancers: Skin – Soft Tissue – Bone – Urogenitals
Dongyou Liu

Tumors and Cancers: Endocrine Glands – Blood – Marrow – Lymph
Dongyou Liu

A Guide to Cancer: Origins and Revelations
Melford John

A Beginner's Guide to Using Open Access Data
Saif Aldeen Saleh AlRyalat and Shaher Momani

Pocket Guide to Bacterial Infections
K. Balamurugan